Systems & Control: Foundations & Applications

Geir E. Dullerud

Control of Uncertain Sampled-Data Systems

Birkhäuser 1996
Boston • Basel • Berlin

Geir E. Dullerud
Department of Electrical Engineering
California Institute of Technology
Pasadena, CA 91125

Library of Congress Cataloging-in-Publication Data

Dullerud, Geir E., 1966-
Control of uncertain sampled-data systems / Geir E. Dullerud.
p. cm. -- (Systems & control)
Includes index.
ISBN 0-8176-3851-2 (hardcover : acid-free)
1. Discrete-time systems. 2. Hybrid computers. 3. Fuzzy systems.
I. Title. II. Series.
QA402.3.D85 1995 95-41966
003'.8--dc20 CIP

Printed on acid-free paper

Birkhäuser

ISBN 0-8176-3851-2
ISBN 3-7643-3851-2
Typeset by the author in TEX
Printed and bound by Quinn-Woodbine, Woodbine, NJ
Printed in the United States of America

9 8 7 6 5 4 3 2 1

To my parents

Contents

1 Introduction **2**
1.1 Modelling and Uncertainty 5
1.2 Summary of Contents 9

2 Preliminaries **12**
2.1 Hilbert Space and Banach Algebras 12
2.2 Operator Theory 15
2.3 Analytic Functions 17
2.4 Time Domain Spaces and Lifting 20
2.5 Frequency Domain Function Spaces 24
2.6 The Structured Singular Value 29

3 Uncertain Sampled-data Systems **31**
3.1 Summary 40

4 Analysis of LTI Uncertainty **41**
4.1 Converting to Frequency Domain 43
4.2 Destabilizing Perturbations 46
4.3 Robustness Test 53
4.4 Sampled-data Frequency Response 60
4.5 Summary 64

5 A Computational Framework **66**
5.1 Lower Bounds 67
5.2 Upper Bounds 72
5.2.1 Convergence 74
5.2.2 Characterization in Finite Dimensions 82
5.2.3 Evaluating $\overline{M}_n$ 90
5.3 An Algorithm 93

5.4 Example 95
5.5 Summary 98

6 Robust Performance **100**
6.1 Robust Performance Conditions 100
6.1.1 Periodic Perturbations 102
6.1.2 Quasi-Periodic Perturbations 107
6.1.3 Arbitrary Time-Varying Uncertainty 114
6.2 Computational Tools 116
6.2.1 Definition and Properties 117
6.2.2 Reduction to Finite Dimensions 123
6.3 Example Algorithm 128
6.3.1 A Cutting Plane Approach 128
6.3.2 Numerical Example 132
6.4 Minimizing the Scaled Hilbert-Schmidt Norm 137
6.4.1 The Hilbert-Schmidt Norm 137
6.4.2 Scaling the Hilbert-Schmidt Norm and Osborne's Method 139
6.5 Summary 142

A State space for $\tilde{M}$ **144**

B Proof of Proposition 5.4 **148**

C State space for $\tilde{M}_n$ **150**

D State Space for Section 6.2 **153**

E Proof of Lemma 6.10 **160**

F The Hilbert-Schmidt Norm of $E_k\check{M}E_l$ **164**

G The S-Procedure **166**

Bibliography **171**

Preface

My main goal in writing this monograph is to provide a detailed treatment of uncertainty analysis for sampled-data systems in the context of systems control theory. Here, sampled-data system refers to the *hybrid* system formed when continuous time and discrete time systems are interconnected; by uncertainty analysis I mean achievable performance in the presence of worst-case uncertainty and disturbances. The focus of the book is sampled-data systems; however the approach presented is applicable to both standard and sampled-data systems.

The past few years has seen a large surge in research activity centered around creating systematic methods for sampled-data design. The aim of this activity has been to deepen and broaden the, by now, sophisticated viewpoint developed for design of purely continuous time or discrete time systems (e.g. $\mathcal{H}_\infty$ or ℓ_1 optimal synthesis, μ theory) so that it can be applied to the design of sampled-data systems. This research effort has been largely successful, producing both interesting new mathematical tools for control theory, and new methodologies for practical engineering design.

Analysis of *structured* uncertainty is an important objective in control design, because it is a flexible and non-conservative way of analyzing system performance, which is suitable in many engineering design scenarios. For this reason it has been studied extensively in a number of mathematical frameworks since the advent of the structured singular value in the early eighties. In this book both analysis techniques and results are developed to address the main sampled-data analysis problems in the context of structured uncertainty. The impact of several related classes of structured dynamic uncertainty on system performance is studied. The resulting development is of both engineering and theoretical interest, and I have endeavored to give a complete picture of which of the book's results are similar to those for purely continuous time systems, and which are surprising and unique to sampled-data systems.

The theory is developed primarily in an operator theoretic framework, and specifically the signal space of interest is $\mathcal{L}_2$. Although all the required background mathematics is collected in Chapter 2, having some familiarity with operators on Hilbert space and frequency domain methods will be a definite asset.

The bulk of this research was carried out from October 1990 to January 1994 while I was a graduate student at Cambridge University. I gratefully acknowledge my sources of financial support during this period which were Peterhouse and the Science and Engineering Research Council (UK). I also acknowledge my current financial support, while a Research Fellow at Caltech, which is provided by the Air Force Office of Scientific Research (USA).

It is a pleasure to thank Keith Glover, my Ph.D. supervisor at Cambridge, for his insightful advice during numerous discussions on the research presented here. I would also like to thank John Doyle for his input and comments, specifically on Chapter 5. Thanks also to Sanjay Lall who read early versions of this manuscript and offered many helpful comments, and Carolyn Beck who did the same with later drafts.

Pasadena, California — Geir E. Dullerud
August, 1995

List of Figures

1.1 General Sampled-data Arrangement 2
1.2 Design Approaches 4

3.1 Uncertain Sampled-data System 32
3.2 Lifted Sampled-data System 36
3.3 Signal Spaces 37
3.4 LTI System 38

4.1 Robust Stabilization Model 42
4.2 Relationships Between the Spaces 44
4.3 Stability Configuration 55

5.1 Modified Configuration 74
5.2 Linear Fractional Map 85
5.3 Loop-shifted System 86
5.4 Additive Perturbation 95
5.5 Upper and Lower Bounds for $\mu_{\Delta_{LTI}}(M(e^{j\omega}))$ with $a = 2$ 97
5.6 Upper and Lower Bounds for $\mu_{\Delta_{LTI}}(M(e^{j\omega}))$ with $a = 100$ . . 98

6.1 Robust Performance Configuration 101
6.2 System Configuration 133
6.3 Plots of $\|\check{M}(e^{j\omega})\|$ and $\bar{\sigma}(M_d(e^{j\omega}))$ 134
6.4 Graphs of $\|D^{\frac{1}{2}}\check{M}(e^{j\omega})D^{-\frac{1}{2}}\|$ and $\bar{\sigma}(D^{\frac{1}{2}}M_d(e^{j\omega})D^{-\frac{1}{2}})$ 135
6.5 Plots of $(d_2/d_1)^{\frac{1}{2}}$ with $\epsilon_{tol} = \{0.005, 10^{-4}\}$ 136

Notation

For the purposes of reference, the following glossary outlines the notation and frequently used objects in the text.

$\mathbb{Z}_+$	the non-negative integers
$\mathbb{R}$, $\mathbb{R}_+$	real and non-negative real numbers
$\mathbb{C}$, $\mathbb{C}_+$, $\bar{\mathbb{C}}_+$	complex number, open right half-plane, closed right half-plane
$\mathbb{D}$, $\bar{\mathbb{D}}$, $\partial\mathbb{D}$	open unit disc, closed unit disc, unit circle
$\cap$	set intersection
$\times$	Cartesian product
$\oplus$	orthogonal direct sum
$\lvert \cdot \rvert_2$, $\langle \cdot, \cdot \rangle$	Euclidean norm and inner product
$*$	conjugate or adjoint
$\lVert \cdot \rVert_{\mathcal{H}}$, $\langle \cdot, \cdot \rangle_{\mathcal{H}}$	norm and inner product on $\mathcal{H}$
$\lVert \cdot \rVert_{\mathcal{H}_1 \to \mathcal{H}_2}$	induced $\mathcal{H}_1$ to $\mathcal{H}_2$ operator norm
$\mathfrak{L}(\mathcal{H}_1, \mathcal{H}_2)$	bounded linear operators from $\mathcal{H}_1$ to $\mathcal{H}_2$
$\mathfrak{C}(\mathcal{H}_1, \mathcal{H}_2)$	compact operators from $\mathcal{H}_1$ to $\mathcal{H}_2$
$\mathcal{U}\mathcal{W}$	open unit ball of the normed space $\mathcal{W}$
$\text{spec}(\cdot)$	spectrum
$\text{rad}(\cdot)$	spectral radius
$\to$	maps to, tends to
$:=$	right hand side defines the left hand side
Im	image of operator
ker	kernel of operator
$\mathcal{H}^\perp$	orthogonal complement of the subspace $\mathcal{H}$
$\bar{\sigma}(\cdot)$	maximum singular value
$\lambda_{\max}(\cdot)$	maximum eigenvalue
$\mathcal{F}_u(\cdot, \cdot)$, $\mathcal{F}_l(\cdot, \cdot)$	upper and lower linear fractional transformations
$\text{diag}(\cdot)$	diagonal operator with given entries
$\mu_{\mathbf{\Delta}}(\cdot)$	structured singular value with respect to the set $\mathbf{\Delta}$
RHS, LHS	right hand and left hand sides

LTI	linear time-invariant
LMI	linear matrix inequality
$\mathcal{L}_2^m$	square integrable functions mapping $[0, \infty)$ to $\mathbb{C}^m$
$\mathcal{K}_2^m$	square integrable functions from $[0, h)$ to $\mathbb{C}^m$
ℓ_2^m	square summable sequences mapping $\mathbb{Z}_+$ to $\mathbb{C}^m$
$\boldsymbol{\ell}_2^m$	square summable sequences mapping $\mathbb{Z}_+$ to $\mathcal{K}_2^m$
$\boldsymbol{\mathcal{H}}_2^m$	square integrable functions from $\mathbb{D}$ to $\mathcal{K}_2^m$
$\boldsymbol{\mathcal{H}}_\infty$, $\boldsymbol{\mathcal{A}}$	spaces of operator valued analytic functions, Section 2.5
$\mathcal{H}_\infty$, $\mathcal{A}_\mathbb{R}$	spaces of matrix valued analytic functions, Section 2.5
$\mathfrak{L}_{\mathcal{A}_\mathbb{R}}$	LTI operators on $\mathcal{L}_2$ defined by $\mathcal{A}_\mathbb{R}$
$\mathfrak{L}_{\boldsymbol{\mathcal{A}}}$	periodic operators on $\mathcal{L}_2$ defined by $\boldsymbol{\mathcal{A}}$
$\boldsymbol{\mathcal{A}}_{\mathcal{A}_\mathbb{R}}$	image of the functions $\mathcal{A}_\mathbb{R}$ in $\boldsymbol{\mathcal{A}}$
$\mathfrak{X}_s$	spatially structured operators on $\mathcal{L}_2$, Chapter 3
$\mathcal{X}$	structured set of matrices, Chapter 4
$\mathcal{A}_\mathbb{R}^\mathcal{X}$	subclass of $\mathcal{A}_\mathbb{R}$ mapping to $\mathcal{X}$
$\mathfrak{X}_{LTI}$	LTI operators in $\mathfrak{X}_s$
$\mathfrak{X}_{PTV}$	intersection of $\mathfrak{X}_s$ and the periodic operators $\mathfrak{L}_{\boldsymbol{\mathcal{A}}}$
$\boldsymbol{\Delta}_{LTI}$	structured set of operators on ℓ_2, Section 4.3
$\boldsymbol{\Delta}_{PTV}$	structured set of operators on $\mathcal{K}_2$, Section 6.1
h	sampling period
$\mathbf{S}$	sampling operator
$\mathbf{H}$	hold operator
W	sampled-data lifting operator, Section 2.4
Z	the z-transform, Section 2.5
$\tilde{U}$	the unilateral shift on $\boldsymbol{\ell}_2$
$\mathbf{D}_h$	the h delay on $\mathcal{L}_2$
ν_k	the sequence $\{0, 1, -1, 2, -2, \ldots\}$

Chapter 1

Introduction

This monograph is concerned with the analysis of uncertain sampled-data systems which arise in control theory.

Dynamical systems encountered in engineering generally evolve in continuous time. Contrasting this, in feedback control the vast majority of complex systems

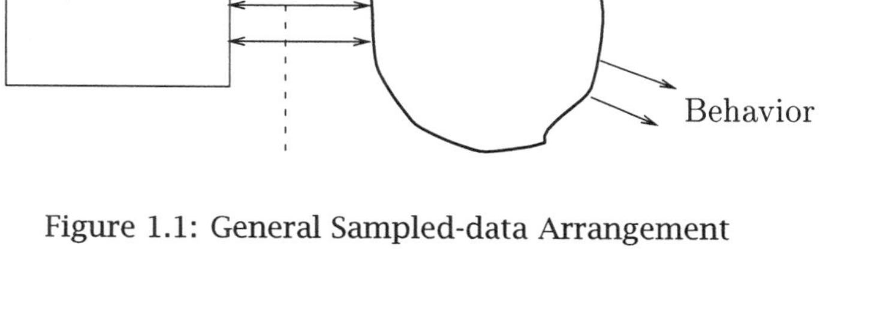

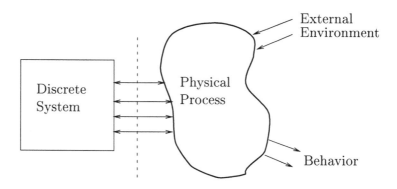

Figure 1.1: General Sampled-data Arrangement

are controlled using *sampled* observations of system behavior taken at discrete time instants. Thus the resulting controlled systems are hybrid, consisting of interacting discrete and continuous components as depicted in

Figure 1.1. Such configurations arise for two main reasons: they are primarily due to the widespread use of digital hardware in feedback control designs; that is, digital computers only have the capability of processing discrete data. Also, some systems only permit discrete measurements; notably, for some chemical and biomedical systems the only sensing instruments available produce discrete signals. These hybrid systems, in which the system to be controlled evolves in continuous time and the controller evolves in discrete time, are called *sampled-data systems.*

The prevailing approach for control design of complex systems is that of model based control. Namely, a mathematical model of the physical system is developed to more easily facilitate study, and thus, a substantial portion of the control system analysis and design is conducted based on the model. This book concentrates exclusively on this particular aspect of control design, where the properties of the mathematical system description are the main focus.

The significant feature of sampled-data system design which distinguishes it from standard techniques for control system design is that it must contend with plant models and control laws lying in different domains. There are three major methodologies for design and analysis of sampled-data systems which are pictorially represented in Figure 1.2, where G is a continuous time process and K_d is a discrete control law. All three methods begin with the first principles continuous time model G and aim to design the discrete time controller K_d and analyze its performance.

The two traditional sampled-data approaches follow the paths around the perimeter of the diagram. The first is to conduct all analysis and design in the continuous time domain using a system that is believed to be a close approximation to the sampled-data system. This is accomplished by associating every continuous time controller K with a discrete time approximation K_d via a discretization method; synthesis and analysis of a controller are then performed in continuous time, with the underlying assumption that the closed loop system behavior obtained with controller K closely reflects that achieved with the sampled-data implementation K_d.

Thus this method does not directly address the issue of implementation in the design stage. The second approach starts instead by discretizing the continuous time system G, giving a discrete time approximation G_d, thus ignoring intersample system behavior. Then the controller K_d is designed directly in discrete time using G_d, with the belief that the performance of this purely discrete system approximates that of the sampled-data system.

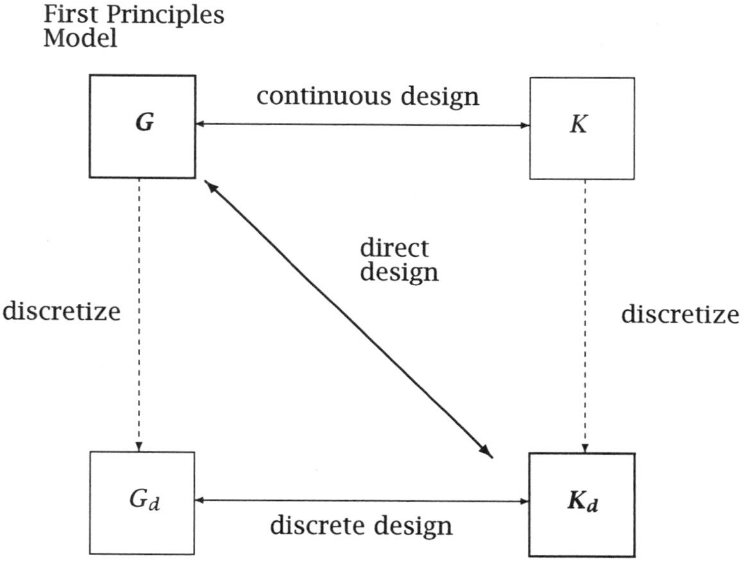

Figure 1.2: Design Approaches

Various design heuristics exist for both of the above traditional approaches and have considerable practical merit in many cases. However, these heuristics amount to approximations and cannot provide precise conclusions about actual hybrid system performance. Recently therefore (see the bibliographic references), there has been considerable research activity devoted to the exact analysis of sampled-data systems where the system G and controller K_d interconnection is treated directly and exactly (shown by the diagonal in Figure 1.2). This research effort into direct design has mainly concentrated on *nominal* stability and performance of sampled-data systems; during the course of this book we will focus on developing a systematic and mathematically precise approach for considering the central design issues of performance and stabilization in the presence of model uncertainty.

1.1 Modelling and Uncertainty

A fundamental barrier limiting the performance of any systematic control design is incomplete knowledge or *uncertainty* about the dynamical features of the system to be controlled. Because of this uncertainty, and the fact that we do not have a complete mathematical representation of the physical world, we can never hope to model all the dynamical aspects of a real system. In feedback control of high performance systems, model uncertainty is especially relevant because control strategies necessarily exploit all available system information to meet demanding performance specifications. For example, a controller synthesis that does not adequately take uncertainty into account may attempt to use system dynamics which are not present. Conversely, it is equally important that system uncertainty not be overly emphasized; otherwise, system performance may be needlessly limited.

The approach used in robust control to address these conflicting requirements is to explicitly include uncertainty in the model description. This is accomplished by using a system model that consists of a set of models: each member of this set represents potential dynamical characteristics of the actual system which cannot be ruled out as impossible or insignificant. A principal method of generating such a mathematical description is to consider a single nominal model together with a set of *perturbations*; these perturbations are applied to the nominal model, thus characterizing a model set. The degree of dynamical system uncertainty can then be realistically and predictably captured by judiciously choosing the *structure* of this perturbation set.

The primary concern of this work is analyzing the effects of structured uncertainty on the performance and stabilization of sampled-data systems which arise in feedback control. The main application of this analysis is therefore evaluation of control strategies and attainable system performance. The general context in which we work is the robustness framework proposed by Zames in [73]; our sampled-data system is comprised of a number of input-output operators on the space of finite energy signals $\mathcal{L}_2$. The nominal sampled-data system model we use is a linear fractional one in which the linear time-invariant plant and controller are connected via sample and hold devices. The structured uncertainty description under consideration is that of dynamic perturbations to the nominal system which enter in a linear fractional way. This uncertainty configuration is motivated by the research of Doyle [20] and co-workers. Four classes of structured causal dynamic uncertainty are examined: linear time-invariant, linear periodically time-varying, quasi-periodic perturbations and arbitrarily time-varying. We make use of a sampled-data lifting technique developed in Bamieh, Pearson, Francis, and Tannenbaum [8], and further develop the frequency domain aspect of this formalism. Similar sampled-data lifting procedures to those in [8] are also developed in Toivonen [66], and Yamamoto [70]. These approaches are all similar to a well-known method in multi-rate signal processing (e.g.[43]), with the important difference that the sampled-data lift results in a system with infinite dimensional inputs and outputs. The basic idea of lifting a periodic system can also be seen in the early work of Kalman and Bertram [36].

A significant portion of this monograph is devoted to analyzing robust stabilization of *linear time-invariant* (LTI) perturbations. This perturbation class arises frequently from the engineering motivation of many design problems, and furthermore is a natural uncertainty set since the nominal plant models considered here are LTI. The robust stabilization problem for sampled-data systems with unstructured LTI uncertainty has been previously considered in Thompson, Athans and Stein [65] and Thompsom, Dailey and Doyle [64], where conic sector analysis was used to obtain sufficient conditions for robust stability. Similarly, Hara, Nakajima and Kabamba [33] used the small gain theorem to also obtain a sufficient condition. Related to these small gain theorem results, Sivashankar and Khargonekar [59] demonstrated that the small gain theorem is a necessary and sufficient test for robust stability if the perturbation class is *expanded* to allow perturbations that are periodic in the nominal system sampling rate.

We develop a condition that exactly characterizes robust stability to structured, and thus also unstructured, LTI perturbations. This robustness test is in terms of the structured singular value[1] of an infinite dimensional compact operator. Note that even in the case of *unstructured* LTI uncertainty, this stability condition is a nontrivial structured singular value problem. For this reason, the small gain condition is not exact and can be extremely conservative when used as a robustness test for LTI perturbations; a simple example which illustrates this point is presented in Chapter 5.

To construct the LTI stability conditions, a special decomposition of sampled-data systems is developed called the *frequency response*, which connects the continuous time frequency domain of the plant and the discrete frequency domain of the controller; this decomposition has an intuitive interpretation and may therefore find wider application. This representation was also independently constructed in a different context by Araki et al. [1, 2].

The sampled-data robust stability condition for LTI uncertainty is particularly challenging to compute because of the aforementioned structure, and this is not totally unexpected. It is by now well-established for purely continuous time LTI systems that the more time variation allowed in the uncertainty set, the simpler the resulting robustness test: if the perturbation set is allowed to include arbitrary time varying perturbations, then an exact robustness test can be computed from a finite dimensional quasi-convex optimization problem [58, 41]; however, for an LTI plant with a structured LTI uncertainty set, the robustness criterion is a structured

[1] defined for matrices in [20]

singular value condition [20, 56], which is considerably more difficult to compute. In the system configurations of this work, the class of *nominal closed-loop* systems considered has been expanded to include sampled-data systems, while the uncertainty set remains LTI; hence the resulting robustness criterion is necessarily more complex than for purely continuous time LTI systems. In the sequel we indeed show that allowing the perturbation class to be time-varying produces simpler robustness conditions for sampled-data systems as well.

Because of the infinite dimensional nature of the exact LTI robustness test above, the approach we adopt to address its evaluation is to develop upper and lower bounds on the stability radius of the sampled-data system. The bounds developed are in terms of two finite dimensional structured singular value problems. These bounds can be systematically improved to enclose the stability radius of the system to any desired degree of accuracy. Thus the engineer can opt for a both known and improved accuracy in the stability radius estimate at the cost of increased computational effort.

In the second main component of this work a different type of robustness problem is examined: we consider *robust performance* of sampled-data systems to structured perturbations, with the important difference being that the uncertainty classes considered have been expanded to allow different amounts of time variation. Solutions to these robust performance problems are pursued using the same operator theoretic framework and machinery that was developed for the LTI robustness analysis. The main feature of each of these conditions is that they are finite dimensional.

The first of these structured uncertainty classes considered consists of perturbations which are periodic in the system sampling rate. The resulting robust performance condition is in terms of a structured singular value problem which has a particularly simple structure compared with the earlier LTI analysis; although it does not in general reduce to a convex computation, it has a finite dimensional form. The other two perturbation classes produce robust performance conditions that are both quasi-convex and finite dimensional; the classes considered are quasi-periodic perturbations and arbitrarily time-varying perturbations. We obtain these results following recent work of Poolla and Tikku [50] on purely discrete time systems. The robust performance conditions we develop can both be evaluated from quasi-convex optimization problems on Euclidean space; these optimization problems have close connections to generalized eigenvalue problems with linear matrix inequality (LMI) constraints. Although the emphasis here is not on computational methods, we develop the main computational tools necessary for evaluating the above quasi-convex programs, and demonstrate their application with a numerical example.

Related to the robust performance problems just described, an additional sufficient condition is also developed which has several computational advantages. This condition is given by the minimum of a set of weighted Hilbert-Schmidt norms of the sampled-data transfer function; we show that this minimization can be converted to a well-known matrix preconditioning problem, considered by Osborne [46], for which an extremely efficient computational method exists.

1.2 Summary of Contents

Chapter 2: Preliminaries

Chapter 2 assembles the various mathematical tools which are required and referenced in the main body of the text. It is divided into six sections: the first three sections concentrate on introducing background mathematics on which subsequent analysis is based. Sections 2.4 and 2.5 introduce and develop the sampled-data lifting formalism that is used extensively throughout the text. Section 2.6 provides a quick introduction to the structured singular value, and gives its generalization to infinite dimensional operators.

Chapter 3: Uncertain Sampled-data Systems

We focus on the problem formulation and developing a unified setting in which to examine robustness issues. The general sampled-data system configuration is introduced; and this periodically time-varying system is lifted to arrive at a time-invariant system with expanded input and output spaces. Robustness of the initial sampled-data system is then recast in this framework.

Chapter 4: Analysis of LTI Uncertainty

The problem of robust stabilization is considered in this chapter which contains four sections. Section 4.1 takes the operator theoretic formulation of the robustness problem and converts it to a sampled-data frequency domain stability test. Using this reformulation, Section 4.2 studies and characterizes the mapping of the LTI perturbation class into the sampled-data frequency domain. Section 4.3 applies the results of the previous sections to obtain an exact characterization of robust stabilization with respect to LTI perturbations; this characterization is stated in terms of a structured singular value condition on the operator-valued sampled-data frequency response. The final section of the chapter develops some of the continuity properties of the sampled-data frequency response of Section 4.3.

Chapter 5: Computational Framework

The main aim is to create a computational framework that exploits the exact stability condition of Chapter 4. The approach is to construct bounds: Section 5.1 derives a sequence of converging lower bounds for this structured singular value. In Section 5.2 an upper bound is developed; this upper bound is considerably more difficult to construct and the section is therefore divided into three subsections that each address aspects of the construction. Section 5.3 combines the derived bounds into an analysis algorithm. In Section 5.4 a numerical example is presented to illustrate the computations of the chapter.

Chapter 6: Robust Performance

Section 6.1 formulates the robust performance problem, and investigates it with respect to three uncertainty classes: periodic, quasi-periodic and arbitrary time-varying uncertainties. Exact conditions for robust performance are demonstrated for each of these uncertainty classes. Section 6.2 develops computational tools for evaluation of the previous robust performance conditions. It is shown that two of the tests can be evaluated by optimization on Euclidean space. Section 6.3 presents an example algorithm to compute robust performance in terms of a quasi-convex optimization problem; a numerical example is provided to illustrate the technique. Section 6.4 is concerned with minimizing the weighted Hilbert-Schmidt norm of the sampled-data transfer function.

Chapter 2

Preliminaries

This chapter is devoted to collecting the mathematical facts and tools required subsequently. The first three sections cover primarily standard results from linear analysis, whereas the remaining three sections provide technical background specific to the analysis of uncertain sampled-data systems.

2.1 Hilbert Space and Banach Algebras

This section identifies the basic underlying structure of all the signal and system spaces encountered later. The material presented here is standard; see for example [11] for more detail on the concepts and definitions given below. Let $\mathcal{W}$ be a vector space over the complex numbers $\mathbb{C}$. A norm on $\mathcal{W}$ is a mapping from $\mathcal{W}$ into the real numbers $\mathbb{R}$ that satisfies,

(i) $\|x\|_{\mathcal{W}} > 0$ for all $x \in \mathcal{W}$ and $x \neq 0$

(ii) $\|\lambda x\|_{\mathcal{W}} = |\lambda| \, \|x\|_{\mathcal{W}}$ for all scalars $\lambda \in \mathbb{C}$ and $x \in \mathcal{W}$

(iii) $\|x + y\|_{\mathcal{W}} \leq \|x\|_{\mathcal{W}} + \|y\|_{\mathcal{W}}$ for all $x, y \in \mathcal{W}$.

A vector space $\mathcal{W}$ together with a norm $\|\cdot\|_{\mathcal{W}}$ is called a *normed space*. When the norm on a space is clear, we usually suppress the subscript on the norm symbol.

In the normed space $\mathcal{W}$, a sequence x_k is defined to converge to x if $\|x_k - x\| \stackrel{k\to\infty}{\longrightarrow} 0$. The normed space $\mathcal{W}$ is *complete* if every Cauchy sequence in $\mathcal{W}$ converges. That is, if x_k is a sequence in $\mathcal{W}$ so that $\|x_k - x_l\| \to 0$ as k and l tend to infinity independently, then x_k converges to some element of $\mathcal{W}$. A complete normed space is called a *Banach space*. A subspace $\mathcal{W}_1$ of a normed space $\mathcal{W}$ is closed if every sequence in it that converges in $\mathcal{W}$ converges to an element in $\mathcal{W}_1$; a closed subspace of a Banach space is therefore complete.

An inner product on a vector space $\mathcal{W}$ is a mapping from the Cartesian product $\mathcal{W} \times \mathcal{W}$ to the complex numbers $\mathbb{C}$, such that for all $x, y, z \in \mathcal{W}$ and each $\lambda \in \mathbb{C}$,

(i) $\langle x, y\rangle_{\mathcal{W}} = \langle y, x\rangle_{\mathcal{W}}^*$

(ii) $\langle x, \lambda y\rangle_{\mathcal{W}} = \lambda\langle x, y\rangle_{\mathcal{W}}$

(iii) $\langle x + y, z\rangle_{\mathcal{W}} = \langle x, z\rangle_{\mathcal{W}} + \langle y, z\rangle_{\mathcal{W}}$

(iv) $\langle x, x\rangle_{\mathcal{W}} > 0$ when $x \neq 0$.

Here $*$ denotes complex conjugation. An inner product induces a norm via $\|x\|^2 = \langle x, x\rangle$; thus a vector space with an inner product, an inner product space, is also a normed space. If an inner product space is complete with respect to this norm, it is by definition a *Hilbert space*.

Given a subspace $\mathcal{H}_1$ of the Hilbert space $\mathcal{H}$, we define its orthogonal complement $\mathcal{H}_1^\perp$ as the subspace

$$\mathcal{H}_1^\perp := \{x \in \mathcal{H} : \langle x, y\rangle = 0 \text{ for all } y \in \mathcal{H}_1\}.$$

If $\mathcal{H}_1$ is a closed subspace, then for each $x \in \mathcal{H}$, there is a unique decomposition $x = y_1 + y_2$ where $y_1 \in \mathcal{H}_1$ and $y_2 \in \mathcal{H}_1^\perp$.

A sequence e_k in a Hilbert space $\mathcal{H}$ is orthonormal if $\|e_k\| = 1$ and $\langle e_k, e_l\rangle = 0$ for all k and l with $k \neq l$. Such a sequence is a *complete orthonormal basis* for $\mathcal{H}$ if the only member of $\mathcal{H}$ that is orthogonal to all e_k is the zero element. That is, if x in $\mathcal{H}$ and $\langle x, e_k\rangle = 0$ for all k, then $x = 0$. The main application of such a basis is that for any $x \in \mathcal{H}$,

$$x = \sum_{k=0}^{\infty} \langle e_k, x\rangle e_k.$$

Furthermore, if α_k is a sequence in $\mathbb{C}$ satisfying $\sum_{k=0}^{\infty} |\alpha_k|^2 < \infty$, then $\sum_{k=0}^{\infty} \alpha_k e_k$ is in $\mathcal{H}$. If x and y are in $\mathcal{H}$ with expansions $x = \sum_{k=0}^{\infty} \alpha_k e_k$ and $y = \sum_{k=0}^{\infty} \beta_k e_k$ then

$$\langle x, y\rangle = \sum_{k=0}^{\infty} \alpha_k^* \beta_k.$$

We will only be using *separable* Hilbert spaces: a Hilbert space $\mathcal{H}$ is separable if it contains a complete orthonormal basis that is countable.

A mapping $U : \mathcal{H}_1 \to \mathcal{H}_2$ where $\mathcal{H}_1$ and $\mathcal{H}_2$ are Hilbert spaces is linear if, for all $x, y \in \mathcal{H}_1$ and all $\alpha, \beta \in \mathbb{C}$, we have $U(\alpha x + \beta y) = \alpha Ux + \beta Uy$. A linear mapping U that is bijective and satisfies

$$\langle x, y\rangle_{\mathcal{H}_1} = \langle Ux, Uy\rangle_{\mathcal{H}_2}$$

for all $x, y \in \mathcal{H}_1$ is by definition an *isomorphism*. If there exists an isomorphism between two Hilbert spaces $\mathcal{H}_1$ and $\mathcal{H}_2$ they are said to be *isomorphic*.

We define the orthogonal direct sum $\mathcal{H}_1 \oplus \mathcal{H}_2$ of two Hilbert spaces to be the space $\mathcal{H}_1 \times \mathcal{H}_2$ with the inner product

$$\langle z_1, z_2\rangle_{\mathcal{H}_1 \oplus \mathcal{H}_2} := \langle x_1, x_2\rangle_{\mathcal{H}_1} + \langle y_1, y_2\rangle_{\mathcal{H}_2}$$

for $z_1 = (x_1, y_1)$ and $z_2 = (x_2, y_2)$ in $\mathcal{H}_1 \times \mathcal{H}_2$.

The final spaces we defined in this section are the *Banach algebras.* Let $\mathcal{B}$ be a Banach space that additionally has a multiplication operation: a multiplication operation is a mapping $\mathcal{B} \times \mathcal{B} \to \mathcal{B}$ satisfying the usual associative property, distributive property with respect to addition, and commutes with scalar multiplication. If in addition to this we have, for all $x, y \in \mathcal{B}$, that the submultiplicative inequality

$$\|xy\| \leq \|x\| \, \|y\|$$

holds, then $\mathcal{B}$ is a Banach algebra.

We have made our definitions commencing with complex vector spaces, and have therefore arrived at complex Hilbert spaces and Banach algebras; the real versions of these spaces are obtained by starting with a real vector space instead.

2.2 Operator Theory

We primarily work with systems that have Hilbert space valued inputs and outputs throughout the sequel. For this reason our goal here is to review some of the important properties of bounded linear operators on a Hilbert space. An excellent reference for more detailed information is [31].

Given two Hilbert spaces $\mathcal{H}_1$ and $\mathcal{H}_2$, a linear mapping $T : \mathcal{H}_1 \to \mathcal{H}_2$ is a bounded linear operator if

$$\|T\|_{\mathcal{H}_1 \to \mathcal{H}_2} := \sup_{\substack{x \in \mathcal{H}_1 \\ x \neq 0}} \frac{\|Tx\|_{\mathcal{H}_2}}{\|x\|_{\mathcal{H}_1}} < \infty.$$

The symbol $\|T\|_{\mathcal{H}_1 \to \mathcal{H}_2}$ signifies the $\mathcal{H}_1$ to $\mathcal{H}_2$ induced norm of T. Bounded linear operators are usually referred to simply as operators. The space of all such operators is denoted by $\mathfrak{L}(\mathcal{H}_1, \mathcal{H}_2)$ and is a Banach space under the induced norm; this notation is abbreviated to $\mathfrak{L}(\mathcal{H}_1)$ when $\mathcal{H}_1$ and $\mathcal{H}_2$ are the same. Multiplication is well-defined on $\mathfrak{L}(\mathcal{H}_1)$ by composition and it is easy to verify that the submultiplicative inequality holds; hence $\mathfrak{L}(\mathcal{H}_1)$ is a Banach algebra.

The above induced norm induces a topology on $\mathfrak{L}(\mathcal{H}_1, \mathcal{H}_2)$, the norm topology; a sequence T_k converges to T in the norm topology if $\|T_k - T\|_{\mathcal{H}_1 \to \mathcal{H}_2} \stackrel{k \to \infty}{\longrightarrow} 0$. Unless otherwise specified, convergence in $\mathfrak{L}(\mathcal{H}_1, \mathcal{H}_2)$ always refers to convergence in the norm topology. Another topology we require is the *weak* topology*: a sequence T_k in $\mathfrak{L}(\mathcal{H}_1, \mathcal{H}_2)$ converges in the weak* topology[1] if there exists a $T \in \mathfrak{L}(\mathcal{H}_1, \mathcal{H}_2)$ so that for each pair

[1] also called the weak operator topology

$x \in \mathcal{H}_1$ and $y \in \mathcal{H}_2$ we have

$$\langle y, T_k x \rangle_{\mathcal{H}_2} \stackrel{k \to \infty}{\longrightarrow} \langle y, Tx \rangle_{\mathcal{H}_2}.$$

This form of convergence in $\mathfrak{L}(\mathcal{H}_1, \mathcal{H}_2)$ is called weak* because $\mathfrak{L}(\mathcal{H}_1, \mathcal{H}_2)$ can be associated with the dual space of another normed space. See for example [19]. A key property of the weak* topology is that every bounded sequence T_k has a subsequence that converges weak* in $\mathfrak{L}(\mathcal{H}_1, \mathcal{H}_2)$.

We denote the adjoint of an operator T by T^*. An operator is self-adjoint if it satisfies $T^* = T$. A self-adjoint operator is defined to be *positive* if

$$\langle x, Tx \rangle_{\mathcal{H}_1} > 0 \quad \text{for all nonzero } x \in \mathcal{H}_1.$$

Similar definitions are made for an operator being negative and nonpositive.

For an operator $T \in \mathfrak{L}(\mathcal{H}_1)$ we say it has an inverse, or is nonsingular, if there exists an element T^{-1} in $\mathfrak{L}(\mathcal{H}_1)$ so that $TT^{-1} = T^{-1}T = I$ the identity map. The *spectrum* of T is the set

$$\text{spec}(T) := \{\lambda \in \mathbb{C} : (I\lambda - T) \text{ is not invertible in } \mathfrak{L}(\mathcal{H}_1)\}.$$

The spectrum is closed and bounded, and we can also define the *spectral radius* function

$$\text{rad}(T) := \max\{|\lambda| : \lambda \in \text{spec}(T)\}.$$

In general $\text{rad}(\cdot)$ is not continuous, but it is upper semicontinuous. The following result provides a bound on the spectral radius.

Proposition 2.1 *Suppose that T is an operator on $\mathcal{H}_1$. Then $rad(T) \leq \|T\|_{\mathcal{H}_1 \to \mathcal{H}_1}$.*

An operator $T \in \mathfrak{L}(\mathcal{H}_1, \mathcal{H}_2)$ is a *compact operator* if for every bounded sequence x_k in $\mathcal{H}_1$, there is a subsequence x_{k_n} so that Tx_{k_n} converges in $\mathcal{H}_2$. We denote the subspace of such operators by $\mathfrak{C}(\mathcal{H}_1, \mathcal{H}_2)$; it is a closed subspace.

We will use the properties of compact operators extensively later, and summarize these below for reference.

Proposition 2.2 *Suppose that T is a compact operator on $\mathcal{H}_1$.*

(i) If $R \in \mathfrak{L}(\mathcal{H}_1)$, then TR is compact.

(ii) The function $rad(\cdot)$ is continuous at T.

(iii) The adjoint operator T^* is compact.

An operator P on $\mathcal{H}_1$ is a projection if $P^2 = P$. A sequence P_k of projections is increasing if, for each x in $\mathcal{H}_1$, the sequence $\|P_k x\|$ is monotonically increasing. Based on such projections we have the following:

Proposition 2.3 *Given $T \in \mathfrak{C}(\mathcal{H}_1)$. If $P_n \in \mathfrak{L}(\mathcal{H}_1)$ is an increasing sequence of projections, so that for every $x \in \mathcal{H}_1$ the sequence $P_n x \overset{n\to\infty}{\longrightarrow} x$, then $P_n T P_n \overset{n\to\infty}{\longrightarrow} T$.*

This result guarantees that T can be approximated, as closely as desired in the norm topology, by matrices on Euclidean space. See [68] for this standard result.

To complete this section we introduce some notation: let Q be a bounded linear operator mapping $\mathcal{H}_1 \oplus \mathcal{H}_2$ to $\mathcal{H}_3 \oplus \mathcal{H}_4$, and $R \in \mathfrak{L}(\mathcal{H}_3, \mathcal{H}_1)$; we conformally partition Q and define the upper linear fractional transformation of the two maps by

$$\mathcal{F}_u(Q, R) := Q_{22} + Q_{21}R(I - Q_{11}R)^{-1}Q_{12},$$

when the inverse of $I - Q_{11}R$ exists. Similarly, when L is in $\mathfrak{L}(\mathcal{H}_4, \mathcal{H}_2)$, we define the lower linear fractional transformation $\mathcal{F}_l(Q, L) := Q_{11} + Q_{12}L(I - Q_{22}L)^{-1}Q_{21}$, if the appropriate inverse exists.

2.3 Analytic Functions

In this short section we generalize the notion of analytic functions and assemble some of their properties. The results can be found in [3].

Let $\mathbb{U}$ be a connected open set in the complex plane and let g be a function mapping $\mathbb{U}$ to a Banach space $\mathcal{B}$. The function is defined to be analytic at $z_0 \in \mathbb{U}$ if the limit

$$\lim_{z \to z_0} \frac{g(z) - g(z_0)}{z - z_0} \quad \text{exists.}$$

It is analytic on $\mathbb{U}$ if it is analytic at each $z \in \mathbb{U}$. Clearly if $g(z)$ is analytic on $\mathbb{U}$ it is also continuous. Note that if $\mathcal{B}$ is the complex plane, the above definition coincides with the usual notion of analyticity. Most of the properties of scalar valued analytic functions carry through to this generalized form. In the sequel we make use of two maximum principles for such functions whose proofs we now outline. They are based on the following mean value property of analytic functions: suppose $z_0 \in \mathbb{U}$ and that $r > 0$ so that all z satisfying $|z - z_0| \leq r$ are in $\mathbb{U}$. Then the analytic function $g(z)$ satisfies

$$g(z_0) = \frac{1}{2\pi j} \int_{-\pi}^{\pi} g(z_0 + re^{j\omega}) d\omega.$$

The integral above is the natural generalization of the Riemann integral to this expanded class of analytic functions, as are all the integrals in the sequel involving operator-valued functions. Hence, from the above mean value property and the triangle inequality, it is easy to see that

$$\|g(z_0)\| \leq \frac{1}{2\pi} \int_{-\pi}^{\pi} \|g(z_0 + re^{j\omega})\| d\omega. \tag{2.1}$$

This integral inequality is a key property of a class of functions called *subharmonic* functions.

A real valued function $q : \mathbb{U} \to \mathbb{R}$ is called subharmonic if it satisfies the following conditions:

(i) $-\infty \leq q(z) < \infty$ for all $z \in \mathbb{U}$.

(ii) q is upper semicontinuous.

(iii) If $z_0 \in \mathbb{U}$ and $r > 0$ so that all z satisfying $|z - z_0| \leq r$ are in $\mathbb{U}$ then

$$q(z_0) \leq \frac{1}{2\pi} \int_{-\pi}^{\pi} q(z_0 + re^{j\omega}) d\omega.$$

(iv) None of the integrals in (iii) is $-\infty$.

Subharmonic functions have many useful properties, some of which we apply below. A main one is that they satisfy a *maximum principle.*

Proposition 2.4 *Suppose q is a subharmonic function on a connected open set $\mathbb{U}$. Then either q has no maximum on $\mathbb{U}$, or q is constant on $\mathbb{U}$.*

We now have the following result which immediately implies a maximum modulus result, via the above proposition, for Banach space valued analytic functions.

Proposition 2.5 *Suppose g is an analytic function mapping a connected open set $\mathbb{U}$ in $\mathbb{C}$ to a Banach space $\mathcal{B}$. Then the real valued function $\|g(z)\|$ is subharmonic on $\mathbb{U}$.*

Proof Since g is analytic, its norm satisfies the inequality in (2.1). Also, the real valued function $\|g(z)\|$ is continuous, and therefore meets all the conditions given above of being subharmonic. Hence, $\|g(z)\|$ satisfies the maximum principle of the claim by invoking Proposition 2.4. ■

Another property of analytic functions is given in the following lemma.

Lemma 2.6 *Suppose g is an analytic function mapping a connected open set $\mathbb{U}$ in $\mathbb{C}$ to a Banach space $\mathcal{B}$. Then* $\log \|g(z)\|$ *is subharmonic on $\mathbb{U}$.*

Later we also require a maximum principle for the spectral radius of an analytic function. The proof of this result is based on the above lemma and the following fact: suppose that T is an operator on the Hilbert space $\mathcal{H}$; then the following equality holds

$$\operatorname{rad}(T) = \lim_{n\to\infty} \|T^n\|^{\frac{1}{n}}$$

and is known as the Gelfand formula for the spectral radius. We can now prove the next proposition.

Proposition 2.7 *Suppose g is an analytic function mapping a connected open set $\mathbb{U}$ in $\mathbb{C}$ to $\mathfrak{L}(\mathcal{H})$, where $\mathcal{H}$ is a Hilbert space. Then rad$(g(z))$ is subharmonic on $\mathbb{U}$.*

Proof Because $g(z)$ is analytic it is routine to show that $g^{2^n}(z)$ is analytic for any $n > 0$. Also, by the preceding lemma we therefore have that each function in the sequence $\frac{1}{2^n}\log\|g^{2^n}(z)\|$ is subharmonic.

Now, by the submultiplicative inequality we have that

$$\|g^{2^{n+1}}(z)\|^{\frac{1}{2^{n+1}}} \leq \|g^{2^n}(z)\|^{\frac{1}{2^n}}$$

holds for each n and $z \in \mathbb{U}$. That is, the sequence $\|g^{2^n}(z)\|^{\frac{1}{2^n}}$ is monotonically decreasing. Therefore the sequence of functions $\frac{1}{2^n}\log\|g^{2^n}(z)\|$ is also decreasing, and each member of the sequence is subharmonic.

By the Gelfand formula we have that

$$\log \operatorname{rad}(g(z)) = \lim_{n\to\infty} \frac{1}{2^n} \log\|g^{2^n}(z)\|.$$

That is, $\log \operatorname{rad}(g(z))$ is the limit of a sequence of monotonically decreasing subharmonic functions. It is an additional property of subharmonicity that such a limit is always subharmonic. So $\log \operatorname{rad}(g(z))$ is subharmonic. See for example [3, App. A].

To prove the claim, another general property of subharmonic functions is used, namely that the composition of a convex and increasing function with a subharmonic one is also subharmonic. Hence, $\operatorname{rad}(g(z))$ is subharmonic since $\log \operatorname{rad}(g(z))$ is and the exponential function is convex and increasing. ■

We now have all the general mathematical concepts we require in the sequel. In the following sections we focus on setting up a formalism specific to sampled-data problems.

2.4 Time Domain Spaces and Lifting

The main objectives here are to introduce the time domain spaces in which our signals will lie, and review a lifting technique for periodic systems presented in [8].

The nominal systems we will consider have their inputs and ouputs in the Hilbert space $\mathcal{L}_2^m$ of functions[2] mapping $[0, +\infty) \to \mathbb{C}^m$,

$$\mathcal{L}_2^m := \{u : \ \|u\|_{\mathcal{L}_2^m}^2 := \int_0^{+\infty} |u(t)|_2^2 \, dt < \infty\},$$

where $|\cdot|_2$ denotes the usual Euclidean norm on $\mathbb{C}^m$. The inner product is given by $\langle u, v\rangle_{\mathcal{L}_2^m} := \int_0^\infty \langle u(t), v(t)\rangle_2 \, dt$, where $\langle \cdot, \cdot \rangle_2$ is the Euclidean inner product on $\mathbb{C}^m$. In discrete time we have the companion Hilbert space of sequences ℓ_2^m mapping the nonnegative integers $\mathbb{Z}_+$ to $\mathbb{C}^m$,

$$\ell_2^m := \{u_d : \|u_d\|_2^2 := \sum_{k=0}^{\infty} |u_d[k]|_2^2 < \infty\},$$

where the inner product is given by $\langle u_d, v_d\rangle_{\ell_2} := \sum_{k=0}^{\infty} \langle u_d[k], v_d[k]\rangle_2$.

At this point we define the sample and hold operators that will appear extensively in later chapters. *Throughout* the sequel, set $h > 0$, the *sampling period*, to be some real number. Then we define the ideal sampler **S** and zero-order hold **H** as the mappings satisfying

$$\begin{aligned} (\mathbf{S}u)[k] &:= u(kh) \\ (\mathbf{H}e_d)(t) &:= e_d[k], \ t \in [kh, \ (k+1)h), \end{aligned} \tag{2.2}$$

for each function $u : [0, +\infty) \to \mathbb{C}^n$ and every sequence $e_d : \mathbb{Z}_+ \to \mathbb{C}^n$. Based on its definition it is easy to see that $\mathbf{H} \in \mathfrak{L}(\ell_2^m, \mathcal{L}_2^m)$ and has norm $\|\mathbf{H}\|_{\ell_2 \to \mathcal{L}_2} = \sqrt{h}$.

Immediately we know that **S** is not in $\mathfrak{L}(\mathcal{L}_2^m, \ell_2^m)$ since it is not well-defined on $\mathcal{L}_2^m$. However, even if we take the space of continuous functions that can be represented in $\mathcal{L}_2^m$, it is easy to show that **S** would still not map them to ℓ_2; a filtering operation is required. Consider the map $\mathbf{F} \in \mathfrak{L}(\mathcal{L}_2^m)$, with $y = \mathbf{F}u$ for $u \in \mathcal{L}_2^m$ that is described by the differential equation

$$\begin{aligned} \dot{x}(t) &= A_F x_F(t) + B_F u(t), \quad x_F(0) = 0 \\ y(t) &= C_F x_F(t), \end{aligned} \tag{2.3}$$

[2]strictly speaking equivalence classes of functions

where A_F is a matrix in $\mathbb{C}^{a\times a}$ with all its eigenvalues having negative real part, $B_F \in \mathbb{C}^{a\times m}$, and $C_F \in \mathbb{C}^{m\times a}$. We define the mapping **SF** to be the map $u \mapsto y_d$ satisfying

$$(SFu)[k] := y_d[k] = C_F \int_0^{kh} e^{A_F(kh-\tau)} B_F u(\tau)\, d\tau$$

from (2.3). Note that the definition of this mapping cannot be generalized to the case where the F has a direct feedthrough term; F must be low-pass. From [17] we have the following proposition[3].

Proposition 2.8 *Given the system defined by (2.3). The mapping* **SF** *is bounded from $\mathcal{L}_2$ to ℓ_2.*

We now move on to notions of time invariance, which play a central role in our framework. Define the delay operator $\mathbf{D}_\tau$ on $\mathcal{L}_2^m$, for each $\tau > 0$, to be the map that takes the function $u(t)$ to the shifted function which is $u(t-\tau)$ for $t \geq \tau$, and zero for $0 \leq t \leq \tau$. An operator $\mathbf{M} \in \mathfrak{L}(\mathcal{L}_2^m)$ is defined to be linear time-invariant (LTI) if $\mathbf{D}_\tau\mathbf{M} = \mathbf{M}\mathbf{D}_\tau$ for all $\tau > 0$. Following the same vein, we define the unilateral shift operator U on ℓ_2^m to be the map $u_d[k] \mapsto u_d[k-1]$, with $(Uu_d)[0] = 0$. A discrete time operator $M_d \in \mathfrak{L}(\ell_2^m)$ is LTI if $M_d U = UM_d$.

In the main part of this work we concentrate on systems which arise from the interconnection of continuous time LTI systems and discrete time LTI systems that are connected via the sample and hold operators **S** and **H**. A simple example of such a system is **HSF** where **F** is from (2.3). However, unless $\mathbf{F} = 0$, there exists $\tau > 0$ so that $\mathbf{D}_\tau\mathbf{HSF} \neq \mathbf{HSFD}_\tau$. That is, the map is not LTI. But it is h-periodic, namely $\mathbf{D}_h\mathbf{HSF} = \mathbf{HSFD}_h$.

The property of being h-periodic is common to all systems formed by discrete time and continuous time LTI systems which are connected through sample and hold operators. The goal of the lifting technique we now describe is to provide a method of representing such systems in a new space where they are LTI.

---

[3]This result is evident from (2.5)

First define the compressed $\mathcal{L}_2^m$ space $\mathcal{K}_2^m$ of functions mapping $[0, h) \to \mathbb{C}^m$ as

$$\mathcal{K}_2^m := \{\psi : \|\psi\|_{\mathcal{K}_2^m}^2 := \int_0^h |\psi(t)|_2^2 \, dt < \infty\},$$

with the inner product $\langle \psi, \phi \rangle_{\mathcal{K}_2^m} := \int_0^h \langle \psi(t), \phi(t) \rangle_2 \, dt$.

Our last time-domain space is denoted by the *boldface* symbol $\boldsymbol{\ell}_2^m$, which is the Hilbert space consisting of functions mapping $\mathbb{Z}_+ \to \mathcal{K}_2^m$:

$$\boldsymbol{\ell}_2^m := \{\tilde{u} : \|\tilde{u}\|_{\boldsymbol{\ell}_2^m}^2 := \sum_{k=0}^{+\infty} \|\tilde{u}[k]\|_{\mathcal{K}_2^m}^2 < \infty\}.$$

We will suppress the dependence of all the above spaces on the dimension m, except in cases where it is of particular relevance.

We now define the sampled-data lifting operator $W : \mathcal{L}_2 \to \boldsymbol{\ell}_2$. For $u \in \mathcal{L}_2$ we have $\tilde{u} = Wu \in \boldsymbol{\ell}_2$ defined by

$$(\tilde{u}[k])(\tau) := u(\tau + kh) \quad \text{for } \tau \in [0, h),\ k \in \mathbb{Z}_+. \tag{2.4}$$

Its inverse, $W^{-1} : \boldsymbol{\ell}_2 \to \mathcal{L}_2$, exists and W is an isomorphism between the spaces $\mathcal{L}_2$ and $\boldsymbol{\ell}_2$. From this it follows that if **F** is a bounded linear operator on $\mathcal{L}_2$, then $\tilde{F} := W\mathbf{F}W^{-1}$ is a bounded linear operator on $\boldsymbol{\ell}_2$. Furthermore,

$$\|\mathbf{F}\|_{\mathcal{L}_2 \to \mathcal{L}_2} = \|\tilde{F}\|_{\boldsymbol{\ell}_2 \to \boldsymbol{\ell}_2}.$$

For finite dimensional LTI operators, the lifted map can be written explicitly by considering the state and output evolution over each time interval $[kh, (k+1)h)$. Suppose **F** is a stable, FDLTI, operator described by (2.3). Then $\tilde{F} := W\mathbf{F}W^{-1}$, for $\tilde{u} \in \boldsymbol{\ell}_2$ and $\tilde{w} = \tilde{F}\tilde{u}$, can be described by

$$\begin{aligned} x_F[k+1] &= \check{A}_F x_F[k] + \check{B}_F \tilde{u}[k] \qquad x_F[0] = 0 \\ \tilde{w}[k] &= \check{C}_F x_F[k] + \check{D}_F \tilde{u}[k], \end{aligned} \tag{2.5}$$

with the definitions given below for $\tau \in [0, h)$

$$\begin{aligned} &\check{A}_F \in \mathbb{C}^{a \times a} && \check{A}_F = e^{A_F h} \\ &\check{B}_F : \mathcal{K}_2 \to \mathbb{C}^a && \check{B}_F \psi = \int_0^h e^{A_F(h-\eta)} B_F \psi(\eta) \, d\eta \\ &\check{C}_F : \mathbb{C}^a \to \mathcal{K}_2 && (\check{C}_F \xi)(\tau) = C_F e^{A_F \tau} \xi \\ &\check{D}_F : \mathcal{K}_2 \to \mathcal{K}_2 && (\check{D}_F \psi)(\tau) = C_F \int_0^\tau e^{A_F(\tau-\eta)} B_F \psi(\eta) \, d\eta. \end{aligned} \tag{2.6}$$

The key property of the isomorphism W is that $\tilde{U}W = W\mathbf{D}_h$, where $\tilde{U}$ is the unilateral shift on ℓ_2; defined as $\tilde{u}[k] \mapsto \tilde{u}[k-1]$, with $(\tilde{U}\tilde{u})[0] = 0$. Hence, if $\mathbf{M}$ is an h-periodic operator on $\mathcal{L}_2$, that is $\mathbf{M}\mathbf{D}_h = \mathbf{D}_h\mathbf{M}$, then the lifted operator $\tilde{M} = W\mathbf{M}W^{-1}$ satisfies

$$\tilde{U}\tilde{M} = \tilde{U}W\mathbf{M}W^{-1} = W\mathbf{D}_h\mathbf{M}W^{-1} = W\mathbf{M}\mathbf{D}_h W^{-1} = \tilde{M}\tilde{U}.$$

So $\tilde{M}$ is linear time-invariant on ℓ_2.

We define an operator $\mathbf{M} \in \mathfrak{L}(\mathcal{L}_2)$ to be *causal* if for every $\tau \geq 0$

$$\mathbf{P}_\tau \mathbf{M}(I - \mathbf{P}_\tau) = 0,$$

where $\mathbf{P}_\tau$ is the truncation map defined on $\mathcal{L}_2$ so that $(\mathbf{P}_\tau u)(t)$ is equal to $u(t)$ for $0 \leq t \leq \tau$ and zero for $t > \tau$. Similarly we define $\tilde{T} \in \mathfrak{L}(\ell_2)$ to be causal if for every $k_0 \in \mathbb{Z}_+$ we have $\tilde{Q}_{k_0}\tilde{T}(I - \tilde{Q}_{k_0}) = 0$ where $\tilde{Q}_{k_0}$ is the truncation map on ℓ_2.

Operators that are LTI and causal have frequency domain representations in terms of analytic functions, that is the topic of the next section.

2.5 Frequency Domain Function Spaces

The frequency domain is now introduced. Our primary references are [27] and [34], which provide additional background to the material presented below.

To begin, we define the normed space $\mathcal{H}_2$, to consist of functions $\check{u}$ mapping $\mathbb{D}$, the open unit disc in $\mathbb{C}$, to elements in $\mathcal{K}_2$ that fulfill the condition

$$\|\check{u}\|_{\mathcal{H}_2}^2 := \sup_{0 \leq r < 1} \frac{1}{2\pi} \int_0^{2\pi} \|\check{u}(re^{j\omega})\|_{\mathcal{K}_2}^2 \, d\omega < \infty.$$

This space is complete and ℓ_2 can be mapped into it by the Gelfand or z-transform: for $\tilde{u} \in \ell_2$ the z-transform $Z : \ell_2 \to \mathcal{H}_2$ is defined by

$$(Z\tilde{u})(z) := \sum_{k=0}^{\infty} \tilde{u}[k] z^k, \tag{2.7}$$

with $z \in \mathbb{D}$. The inverse mapping $Z^{-1} : \mathcal{H}_2 \to \ell_2$ is also well-defined, and we have the next proposition that follows from [63, pp.184–185].

Proposition 2.9 *The spaces ℓ_2 and $\mathcal{H}_2$ are isomorphic via the z-transform pair previously defined.*

We now examine the image of the LTI operators on ℓ_2 when they are mapped by the z-transform to act on $\mathcal{H}_2$. To start, we consider the specific example of the operator $\widetilde{F}$ described by (2.5), with all the eigenvalues of the matrix $\check{A}_F$ in $\mathbb{D}$. Suppose that $\tilde{u} \in \ell_2$ and set $\tilde{w} = \widetilde{F}\tilde{u}$. Then the z-transform of $\tilde{w}$, the function $\check{w}$, is given by

$$\check{w}(z) := \sum_{k=0}^{\infty} \tilde{w}[k]z^k = \sum_{k=1}^{\infty} z^k (\sum_{l=0}^{k-1} \check{C}_F \check{A}_F^{k-l-1} \check{B}_F \tilde{u}[l] + \check{D}_F \tilde{u}[k]) + \check{D}_F \tilde{u}[0],$$

for $z \in \mathbb{D}$. Because $\check{A}_F$ is a stable matrix the above series is absolutely convergent with respect to $\|\cdot\|_{\mathcal{K}_2}$, and we can rearrange the terms to get

$$\begin{aligned} \check{w}(z) &= \left((\sum_{m=0}^{\infty} z^{m+1} \check{C}_F \check{A}_F^m \check{B}_F) + \check{D}_F \right) \sum_{q=0}^{\infty} \tilde{u}[q] z^q \\ &= (\check{C}_F z (I - z\check{A}_F)^{-1} \check{B}_F + \check{D}_F) \check{u}(z) =: \check{F}(z)\check{u}(z), \end{aligned} \tag{2.8}$$

where $\check{u} \in \mathcal{H}_2$ is the z-transform of $\tilde{u}$. Hence, the image of $\widetilde{F}$ is a multiplication operator on $\mathcal{H}_2$. For later reference, note that the definition of the function $\check{F}(z)$ can be extended to $z \in \bar{\mathbb{D}}$, since $\check{A}_F$ is a stable matrix.

Our aim now is to define two classes of multiplication operators on $\mathcal{H}_2$ that generalize the above example. In the sequel we use $\|\cdot\|$ to signify the $\mathcal{K}_2 \to \mathcal{K}_2$ induced norm. Define the normed algebra $\mathcal{H}_\infty$ to consist of functions $\check{G} : \mathbb{D} \to \mathcal{L}(\mathcal{K}_2)$, which are analytic on $\mathbb{D}$ and have finite norm defined by

$$\|\check{G}\|_\infty := \sup_{|z|<1} \|\check{G}(z)\|.$$

In particular, $\check{F}$ in (2.8) is an element of $\mathcal{H}_\infty$.

For each element $\check{G} \in \mathcal{H}_\infty$, we can define a multiplication operator, denoted by $\Theta_{\check{G}}$, on $\mathcal{H}_2$ by

$$(\Theta_{\check{G}} \check{u})(z) := \check{G}(z)\check{u}(z), \tag{2.9}$$

for $\check{u} \in \mathcal{H}_2$ and $z \in \mathbb{D}$. From [27, pp. 234] we have the precise connection between the linear operators $\mathcal{L}(\ell_2)$ and functions $\mathcal{H}_\infty$:

Proposition 2.10

(i) If $\tilde{G} \in \mathfrak{L}(\ell_2)$ *is LTI and causal, then there exists* $\check{G} \in \mathcal{H}_\infty$ *so that* $\tilde{G} = Z^{-1}\Theta_{\check{G}}Z$.

(ii) Every multiplication operator $\Theta_{\check{G}}$*, defined from a function* $\check{G} \in \mathcal{H}_\infty$*, defines a bounded linear operator on* $\mathcal{H}_2$*. Moreover, the* $\mathcal{H}_2 \to \mathcal{H}_2$ *induced norm of the operator is equal to* $\|\check{G}\|_\infty$.

The proposition states that the space of causal LTI operators on ℓ_2 is isomorphic to $\mathcal{H}_\infty$. Furthermore, each such operator on ℓ_2 can be associated with an h-periodic operator on $\mathcal{L}_2$. Thus $\mathcal{H}_\infty$ also provides a representation for a particular subspace of $\mathfrak{L}(\mathcal{L}_2)$.

It is convenient to define a space of functions related to $\mathcal{H}_\infty$ which consists of functions that have continuous extensions to the closed unit disc $\bar{\mathbb{D}}$: let $\mathcal{A}$ be the normed space of functions $\check{G} : \bar{\mathbb{D}} \to \mathfrak{L}(\mathcal{K}_2)$ which are analytic on $\mathbb{D}$, continuous on $\bar{\mathbb{D}}$, along with the norm

$$\|\check{G}\|_\infty := \max_{w \in \mathbb{R}} \|G(e^{j\omega})\| = \max_{z \in \bar{\mathbb{D}}} \|G(z)\|.$$

This is identical to the norm on $\mathcal{H}_\infty$ since functions $\|\check{G}(z)\| : \mathbb{D} \to \mathbb{R}$ satisfy a maximum principle; we therefore regard $\mathcal{A}$ as a subspace of $\mathcal{H}_\infty$. For later reference we will denote the space of operators on $\mathcal{L}_2$ that can be identified with the transfer functions $\mathcal{A}$ by $\mathfrak{L}_{\mathcal{A}}$.

Proposition 2.11 *The spaces* $\mathcal{H}_\infty$ *and* $\mathcal{A}$ *are complete.*

Proof That $\mathcal{H}_\infty$ is complete is standard; see [27].

Since $\mathcal{A}$ is a subspace of $\mathcal{H}_\infty$, it is sufficient to show that every Cauchy sequence converges to a continuous function on $\bar{\mathbb{D}}$. Suppose $\check{T}_k$ is a Cauchy sequence in $\mathcal{A}$. Then by definition of the norm we have for each $z \in \bar{\mathbb{D}}$ that $\|\check{T}_k(z) - \check{T}_l(z)\| \le \|\check{T}_k - \check{T}_l\|_\infty$, which tends to zero as k and l tend to infinity. Hence, T_k tends uniformly to a function $\check{T} : \bar{\mathbb{D}} \to \mathfrak{L}(\mathcal{K}_2)$; since the functions $\check{T}_k$ are continuous so is $\check{T}$. ■

Analogous to these two spaces we define their half-plane matrix valued versions. A function $\hat{\Delta}$ mapping the open right half-plane $\mathbb{C}_+$ to the matrices in $\mathbb{C}^{m\times p}$ is in $\mathcal{H}_\infty$ if it is analytic on $\mathbb{C}_+$ and its norm defined by

$$\|\hat{\Delta}\|_\infty := \sup_{s\in\mathbb{C}_+} \bar{\sigma}(\hat{\Delta}(s)),$$

is finite; the symbol $\bar{\sigma}(\cdot)$ is the maximum singular value. Note that we have also used $\|\cdot\|_\infty$ to denote the norm on $\boldsymbol{\mathcal{H}}_\infty$. The context will always indicate which is meant. We define the subspace of $\mathcal{H}_\infty$, the half-plane algebra $\mathcal{A}_\mathbb{R}$ to be comprised of functions $\hat{\Delta} : \bar{\mathbb{C}}_+ \to \mathbb{C}^{m\times p}$ that are continuous on $\bar{\mathbb{C}}_+ \cup \{\infty\}$, analytic on $\mathbb{C}_+$, with the norm $\|\hat{\Delta}\|_\infty := \sup_{\omega\in\mathbb{R}} \bar{\sigma}(\hat{\Delta}(j\omega))$; we put a further restriction on the functions in $\mathcal{A}_\mathbb{R}$, namely that for each $s \in \bar{\mathbb{C}}_+$ they satisfy

$$\hat{\Delta}(s)^* = \hat{\Delta}(s^*)^T,$$

where $*$ denotes Hermitian conjugate and the superscript ^{T} means matrix transpose.

In a similar way to the definition in (2.9), any function $\hat{\Delta}$ in $\mathcal{H}_\infty$ defines a causal operator $\boldsymbol{\Delta}$ on $\mathcal{L}_2$ via multiplication and the Laplace transform. For reference define the subspace $\mathfrak{L}_{\mathcal{A}_\mathbb{R}} \subset \mathfrak{L}(\mathcal{L}_2)$ of operators that can be represented by transfer functions in $\mathcal{A}_\mathbb{R}$.

Proposition 2.12 *The stable, proper, real rational, transfer functions are dense in $\mathcal{A}_\mathbb{R}$.*

Proof The space $\mathcal{A}_\mathbb{R}$ is isometrically isomorphic to the disc algebra by the bilinear map $z = (1-s)(1+s)^{-1}$, in which the real polynomials are dense. See [35]. ■

An attractive feature of the spaces $\mathcal{A}_\mathbb{R}$ and $\boldsymbol{\mathcal{A}}$ is that their elements are defined on the imaginary axis $j\mathbb{R}$ and the unit circle $\partial\mathbb{D}$, the boundaries of their domains. The functions in $\mathcal{H}_\infty$ and $\boldsymbol{\mathcal{H}}_\infty$ do not share this property, but they almost do:

Proposition 2.13

(i) *Given $\check{\Delta} \in \boldsymbol{\mathcal{H}}_\infty$. Then there exists a set $\mathbb{O} \subset [-\pi, \pi]$ of Lebesgue measure zero so that for every $\psi \in \mathcal{K}_2$ the limit*

$$\lim_{r \to 1} \check{\Delta}(r\, e^{j\omega})\psi \quad \text{exists in } \mathcal{K}_2 \text{ for each } \omega \in [-\pi, \pi] \backslash \mathbb{O}.$$

(ii) *Given $\hat{\Delta} \in \mathcal{H}_\infty$. Then there exists a set $\mathbb{S} \subset \mathbb{R}$ of Lebesgue measure zero so that the limit*

$$\lim_{\sigma \to 0} \hat{\Delta}(\sigma + j\omega) \quad \text{exists for each } \omega \in \mathbb{R} \backslash \mathbb{S}.$$

See for example [27].

Based on the proposition we can extend the domain of definition of $\hat{\Delta} \in \mathcal{H}_\infty$ and $\check{\Delta} \in \boldsymbol{\mathcal{H}}_\infty$ to their respective boundaries

$$\begin{aligned} \hat{\Delta}(j\omega) &:= \lim_{\sigma \to 0} \hat{\Delta}(\sigma + j\omega), \\ \check{\Delta}(e^{j\omega}) &:= \lim_{r \to 1} \check{\Delta}(r\, e^{j\omega}) \end{aligned}$$

where the limits exist. This is consistent with definitions of the subspaces $\mathcal{A}_\mathbb{R}$ and $\boldsymbol{\mathcal{A}}$. Along with the extended definition of functions goes the following result.

Proposition 2.14 *If $\hat{\Delta} \in \mathcal{H}_\infty$ and $\check{\Delta} \in \boldsymbol{\mathcal{H}}_\infty$, then*

$$\begin{aligned} \|\hat{\Delta}\|_\infty &= \operatorname*{ess\ sup}_{\omega \in \mathbb{R}} \bar{\sigma}(\hat{\Delta}(j\omega)), \\ \|\check{\Delta}\|_\infty &= \operatorname*{ess\ sup}_{\omega \in [-\pi, \pi]} \|\check{\Delta}(e^{j\omega})\|. \end{aligned}$$

We have now defined and developed all the tools required in order to describe the dynamic aspects of the systems to be considered. In the next section we will concentrate on the spatial structure of systems.

2.6 The Structured Singular Value

We introduce the structured singular value of a linear operator. The structured singular value was first introduced in [20] for complex matrices; also [56] contains earlier related work. As we will see in the next chapter this is a convenient tool for studying systems with structured uncertainty.

We make the natural generalization of the structured singular value to operators: suppose that $M \in \mathfrak{L}(\mathcal{H}_1, \mathcal{H}_2)$ and $\boldsymbol{\Delta}$ is a subspace of $\mathfrak{L}(\mathcal{H}_2, \mathcal{H}_1)$; then the structured singular value of M with respect to the set $\boldsymbol{\Delta}$ is defined as

$$\mu_{\boldsymbol{\Delta}}(M) := (\inf\{\|\Delta\| : \Delta \in \boldsymbol{\Delta} \text{ and } I - M\Delta \text{ is singular}\})^{-1}.$$

A property of the structured singular value, which is straightforward to verify, is that

$$\mu_{\boldsymbol{\Delta}}(M) = \sup_{\Delta \in \mathcal{U}\boldsymbol{\Delta}} \operatorname{rad}(M\Delta),$$

where $\mathcal{U}\boldsymbol{\Delta}$ denotes the open unit ball of $\boldsymbol{\Delta}$.

When M is a complex matrix, of say dimension $m \times m$, the perturbation set in $\mathbb{C}^{m\times m}$ is usually defined to be of the form

$$\boldsymbol{\Delta} := \{\operatorname{diag}(\delta_1 I_{m_1}, \ldots, \delta_s I_{m_s}, \underline{\Delta}_{s+1}, \ldots, \underline{\Delta}_d) : \delta_i \in \mathbb{C},\ \underline{\Delta}^k \in \mathbb{C}^{m_k \times m_k}\}, \tag{2.10}$$

for a fixed set of positive integers m_k, which must be compatible with m. Even in this case, where $\mathcal{H}_1$ and $\mathcal{H}_2$ are Euclidean spaces, computing $\mu_{\boldsymbol{\Delta}}(M)$ is difficult, and at present there is no solution to this problem in general. For a complete synopsis of the current results on the structured singular value of a matrix, see [49]. Below we give the basic ideas used to tackle estimation of the structured singular value.

Instead of attempting to compute the structured singular value, the standard approach is to develop methods for computing upper and lower bounds for $\mu_{\boldsymbol{\Delta}}(M)$. Let $M \in \mathbb{C}^{m\times m}$ and define the sets $\mathcal{D}_{\boldsymbol{\Delta}}$ and $\mathcal{Q}_{\boldsymbol{\Delta}}$:

$$\begin{aligned} \mathcal{D}_{\boldsymbol{\Delta}} &:= \{D \in \mathbb{C}^{m\times m} : D \text{ nonsingular},\ D\Delta = \Delta D \ \text{ for all } \Delta \in \boldsymbol{\Delta}\} \\ \mathcal{Q}_{\boldsymbol{\Delta}} &:= \{Q \in \boldsymbol{\Delta} : Q^*Q = I\}. \end{aligned}$$

With these definitions we have the following inequalities.

Proposition 2.15 $\sup_{Q \in \mathcal{U}\mathcal{Q}_\Delta} \operatorname{rad}(MQ) \leq \mu_{\mathbf{\Delta}}(M) \leq \inf_{D \in \mathcal{D}_\Delta} \|DMD^{-1}\|$

Proof The LHS inequality is obvious since $\mathcal{Q}_{\mathbf{\Delta}} \subset \mathbf{\Delta}$.

To show the second inequality, choose any $D \in \mathcal{D}_{\mathbf{\Delta}}$ and $\Delta \in \mathbf{\Delta}$, and recall that they commute. The claim follows directly from the inequalities

$$\begin{aligned} \operatorname{rad}(M\Delta) &= \operatorname{rad}(DM\Delta D^{-1}) = \operatorname{rad}(DMD^{-1}\Delta) \\ &\leq \|DMD^{-1}\Delta\| \leq \|DMD^{-1}\| \cdot \|\Delta\| \quad \blacksquare \end{aligned}$$

Considerable research has been done on computing the bounds in Proposition 2.15 when M is a matrix and the uncertainty set is as in (2.10). The lower bound is still not computable exactly, but a reasonable numerical procedure has been developed. See [47]. However, the upper bound can be computed to any desired degree of accuracy; it is a quasi-convex optimization problem.

We are now ready to effectively address robustness issues in the remainder of the text.

Chapter 3

Uncertain Sampled-data Systems

In this chapter we introduce the general sampled-data system configuration considered in this work. The system arrangement is shown in Figure 3.1. The figure shows a continuous time system $\mathbf{G}$ in feedback with a discrete time controller $\mathbf{K}_d$ through the sample and hold devices defined in (2.2). Also in feedback with $\mathbf{G}$ is the block diagonal system $\text{diag}(\boldsymbol{\Delta}^1, \ldots, \boldsymbol{\Delta}^d)$, referred to as simply $\boldsymbol{\Delta}$, which represents uncertainty in the connected system. The system $\mathbf{G}$ is assumed to be a finite dimensional linear time-invariant (FDLTI) continuous time system; the discrete time controller $\mathbf{K}_d$ is also FDLTI.

The input w represents an exogenous signal to the system that contains, for example, disturbances, noise and command signals. The output z is the regulated output that is to be attenuated. In addition to these signals we have p_1 and p_2, the internal inputs, to which the system is to be stable, and the internal outputs q_1 and q_2. The signal y contains the measured outputs of the system, and u the controlled inputs to the system. All the inputs considered are in $\mathcal{L}_2$ the space of finite energy signals.

The configuration in Figure 3.1 is a generalization of a robust performance paradigm for purely continuous time systems to sampled-data systems. This robust performance configuration was first introduced in Doyle [20] for continuous time signals; it can incorporate standard models of uncertainty, such as additive, multiplicative and coprime factor uncertainty,

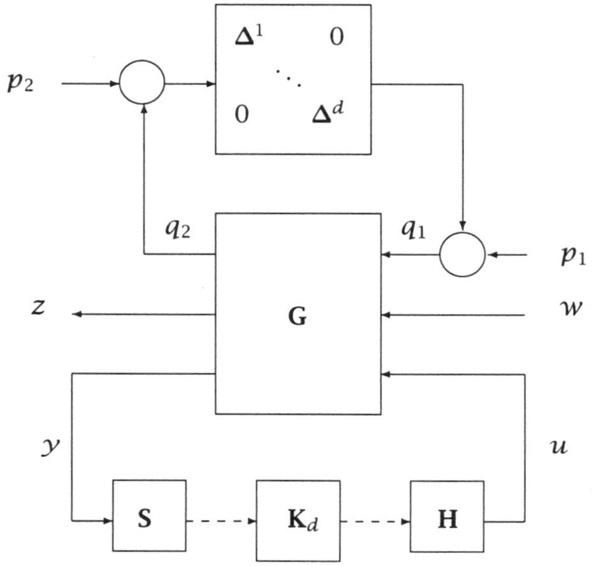

Figure 3.1: Uncertain Sampled-data System

into its structure, as well as a number of other types of uncertainty. A key feature of the block diagonal structure of the uncertainty is that it allows for nonconservative modelling of uncertainty occurring in different parts of a system. See [4] and [47] for further engineering motivation of this uncertainty model.

We now state the precise assumptions on our system components: let (A, B, C, D) be a minimal state space realization for the FDLTI system **G**, where $A \in \mathbb{C}^{n_A \times n_A}$, $B \in \mathbb{C}^{n_A \times (m+r+l_c)}$, $C \in \mathbb{C}^{(m+r+l_o) \times n_A}$ and $D \in \mathbb{C}^{(m+r+l_o)\times(m+r+l_c)}$. The dimension n_A is the size of the state space of the system; l_o is the number of measured outputs and l_c is the number of control channels; m denotes the dimension of the internal signals q_1 and q_2; and r is dimension of the signals w and z. Note that the nominal closed loop system is square; this assumption is used to make the resulting notation as simple as possible. It can be relaxed, and with primarily cosmetic modifications all the results in the sequel hold for general nonsquare systems.

We denote the transfer function of $\mathbf{G}$ by $\hat{G}(s) := C(Is - A)^{-1}B + D$ and adopt the notation

$$\hat{G}(s) =: \left[\begin{array}{c|ccc} A & B_1^1 & B_1^2 & B_2 \\ \hline C_1^1 & 0 & 0 & D_{12}^1 \\ C_1^2 & 0 & 0 & D_{12}^2 \\ C_2 & 0 & 0 & 0 \end{array}\right] =: \begin{bmatrix} \hat{G}_{11} & \hat{G}_{12} \\ \hat{G}_{21} & \hat{G}_{22} \end{bmatrix}, \tag{3.1}$$

where the transfer function has been conformally partitioned with respect to the inputs and outputs of Figure 3.1. Note that D has a special structure: the matrix $D_{21} = 0$ is a necessary assumption so that the signals entering the sampler $\mathbf{S}$ are filtered as required by Proposition 2.8. For simplicity, it is assumed that $D_{22} = 0$ and $D_{11} = 0$; these conditions can be removed without affecting subsequent results.

Also, let $(A_{K_d}, B_{K_d}, C_{K_d}, D_{K_d})$ be a minimal state space realization for the discrete-time controller $\mathbf{K}_d$. The above realizations for $\mathbf{G}$ and $\mathbf{K}_d$ are referred to throughout the sequel.

The perturbation $\mathbf{\Delta}$ to the system is assumed to be a member of the spatially structured set

$$\mathfrak{X}_s := \{\mathbf{\Delta} = \text{diag}(\mathbf{\Delta}^1, \ldots, \mathbf{\Delta}^d) : \mathbf{\Delta}^k \in \mathfrak{L}(\mathcal{L}_2^{m_k}) \text{ for } 1 \le k \le d\}, \tag{3.2}$$

where $\sum_{k=1}^{d} m_k = m$. These perturbations are of the structure shown in Figure 3.1. By spatial we refer to the fact that structure is imposed on the Euclidean part of a member of $\mathfrak{X}_s$. Namely, given $\mathbf{\Delta} = \text{diag}(\mathbf{\Delta}^1, \ldots, \mathbf{\Delta}^d) \in \mathfrak{X}_s$ and $u = (u_1, \ldots, u_d) \in \mathcal{L}_2^m$, with $u_k \in \mathcal{L}_2^{m_k}$, the mapping is given by $\mathbf{\Delta}u = (\mathbf{\Delta}_1 u_1, \ldots, \mathbf{\Delta}_d u_d)$. Thus the sequence m_k of indices defines a spatial partitioning of the uncertainty. For notational simplicity we assume that these spatial blocks are square, however with minor changes this assumption can be removed. The uncertainty sets we will consider in the text are all subsets of the spatial set $\mathfrak{X}_s$, and we will vary the *dynamic* structure of our perturbation set: in Chapters 4 and 5 we consider a subset of LTI operators in $\mathfrak{X}_s$; in Chapter 6 we deal with the set of h-periodic operators, the set of quasi h-periodic operators and the set of causal operators contained in $\mathfrak{X}_s$.

We have a *standing* assumption of nominal stability of the interconnection between the plant $\mathbf{G}$ and controller $\mathbf{K}_d$:

Assumption 3.1 *Suppose in Figure 3.1 that $w = p_1 = p_2 = 0$. Then for any initial states $x_G(0)$ and $x_{K_d}[0]$ of the minimal realizations of $\mathbf{G}$ and $\mathbf{K}_d$ respectively, the limits $x_G(t) \overset{t\to\infty}{\longrightarrow} 0$ and $x_{K_d}[k] \overset{k\to\infty}{\longrightarrow} 0$ are satisfied.*

The above amounts to imposing asymptotic stability on the sampled-data system; this assumption guarantees input-output stability of the nominal system, and is equivalent to it if a mild nonpathological sampling condition is imposed on the sampling rate. See [16]. In the sequel we will see that the nominal stability condition above can be easily verified by examining the eigenvalues of a matrix.

The main objective of this work is to study robust stabilization and performance of the systems defined by Figure 3.1 to different classes of perturbations. We now make the notions of robust stabilization and performance precise. In the definition below, $\mathcal{U}\mathfrak{X}$ denotes the open unit ball of the subspace $\mathfrak{X}$.

Definition 3.2 *Suppose that $\mathfrak{X}$ is a subspace of $\mathfrak{L}(\mathcal{L}_2)$ and $r > 0$. Then the system in Figure 3.1 has* robust stabilization *with respect to perturbations $r\mathcal{U}\mathfrak{X}$ if the map*

$$\begin{bmatrix} w \\ p_1 \\ p_2 \end{bmatrix} \mapsto \begin{bmatrix} z \\ q_1 \\ q_2 \end{bmatrix} \quad \text{exists and is bounded on } \mathcal{L}_2 \text{ for each } \Delta \in r\mathcal{U}\mathfrak{X}.$$

Related to robust stabilization we have robust performance, which is the property of achieving a particular performance in the face of worst-case uncertainty.

Definition 3.3 *Suppose that $\mathfrak{X}$ is a subspace of $\mathfrak{L}(\mathcal{L}_2)$ and $r > 0$. Then the system in Figure 3.1 has* robust performance *with respect to perturbations $r\mathcal{U}\mathfrak{X}$ if it is robustly stabilized and the performance inequality $r\,\|w \mapsto z\|_{\mathcal{L}_2\to\mathcal{L}_2} \leq 1$ is maintained for all $\Delta \in r\mathcal{U}\mathfrak{X}$.*

Here, robust performance has been defined with the same scale r on both the uncertainty set and the performance inequality; note that $\mathbf{G}$ can be scaled a priori so that, in practice, the size of the uncertainty set and bound on the performance inequality can be varied independently. In most cases, for convenience, we set $r = 1$ when stating and proving results. This is done without loss of generality since all the systems we consider are linear, and can therefore be scaled appropriately for general r.

It is these two system properties on which we focus throughout. Chapters 4 and 5 are concerned exclusively with the analysis of robust stabilization, whereas Chapter 6 addresses issues of robust performance.

The first step in analyzing both these properties is to transform the sampled-data system into a more advantageous form. The transformation is motivated by the fact that the nominal feedback system $\mathcal{F}_l(\mathbf{G}, \mathbf{HK}_d\mathbf{S})$ is h-periodic:

$$\mathbf{D}_h\mathcal{F}_l(\mathbf{G}, \mathbf{HK}_d\mathbf{S}) = \mathcal{F}_l(\mathbf{G}, \mathbf{HK}_d\mathbf{S})\mathbf{D}_h. \tag{3.3}$$

The system is however *not* LTI on $\mathcal{L}_2$, in general, because of the periodicity of the sample and hold operators. We therefore apply the lifting formalism of Section 2.4 to obtain an LTI representation of the system on ℓ_2.

The sampled-data system is lifted by augmenting the configuration in Figure 3.1 with the operator W, defined in Section 2.4, as shown in Figure 3.2. The inputs, $\tilde{w}$, $\tilde{p}_1$ and $\tilde{p}_2$, to this new system are now taken in ℓ_2. Because of the isomorphism between $\mathcal{L}_2$ and ℓ_2 the maps $w, p_1, p_2 \mapsto z, q_1, q_2$ in Figure 3.1 are bounded on $\mathcal{L}_2$ if and only if the maps $\tilde{w}, \tilde{p}_1, \tilde{p}_2 \mapsto \tilde{z}, \tilde{q}_1, \tilde{q}_2$ are bounded on ℓ_2.

Referring to Figure 3.2 we set

$$\tilde{\Delta} := W\mathbf{\Delta}W^{-1}$$

and define the operator in the lower dotted box to be $\tilde{M}$:

$$\tilde{M} := W\mathcal{F}_l(\mathbf{G}, \mathbf{HK}_d\mathbf{S})W^{-1}.$$

The operator $\tilde{M}$ is an LTI operator on ℓ_2 as stated in (3.3); it is described

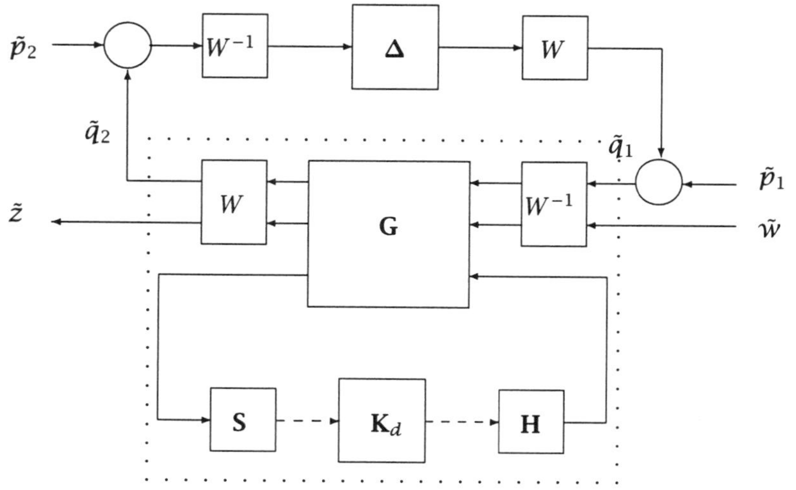

Figure 3.2: Lifted Sampled-data System

by the difference equations

$$\begin{aligned} x[k+1] &= A_d x[k] + \check{B}\tilde{u}[k], \quad x[0] = 0 \\ \tilde{y}[k] &= \check{C}x[k] + \check{D}\tilde{u}[k], \end{aligned} \tag{3.4}$$

where $\tilde{u} \in \ell_2$ and $\tilde{y} = \tilde{M}\tilde{u}$. Here A_d is a matrix in $\mathbb{C}^{\tilde{n}\times\tilde{n}}$, $\check{B} : \mathcal{K}_2 \to \mathbb{C}^{\tilde{n}}$, $\check{C} : \mathbb{C}^{\tilde{n}} \to \mathcal{K}_2$ and $\check{D} : \mathcal{K}_2 \to \mathcal{K}_2$. The form of these operators is similar to that in (2.5), and they are derived by considering the evolution of the sampled-data system over each time interval $[kh, (k+1)h)$. We do not require explicit formulae for these operators at the moment; we use them later. Exact formulae can be found in Appendix A. The state $x[k]$ above corresponds to the states of the plant and controller at the sampling instants. That is,

$$x[k] = \begin{bmatrix} x_G(kh) \\ x_{K_d}[k] \end{bmatrix}$$

for $k \geq 0$, where x_G and x_{K_d} are the states of the minimal realizations for **G** and $\mathbf{K}_d$. Hence, by Assumption 3.1 the matrix A_d in (3.4) must have its eigenvalues in the open unit disc $\mathbb{D}$.

From the representation of $\tilde{M}$ in (3.4) it is clear that $\tilde{M}$ is LTI, that is

$$\tilde{U}\tilde{M} = \tilde{M}\tilde{U},$$

where $\tilde{U}$ is the unilateral shift, and therefore by Proposition 2.10 it can be associated with a transfer function in $\mathcal{H}_\infty$. By the same procedure as in (2.8) it is straightforward to verify that this transfer function is

$$\check{M}(z) := \check{C}z(Iz - A_d)^{-1}\check{B} + \check{D}. \tag{3.5}$$

As noted above, all the eigenvalues of A_d have modulus less than one and therefore

$$\check{M} \in \mathcal{A}.$$

By the isomorphism in Proposition 2.10 we know that $\|\tilde{M}\|_{\ell_2 \to \ell_2} = \|\check{M}\|_\infty$, and therefore by the fact that W is an isomorphism that

$$\|\mathbf{M}\|_{\mathcal{L}_2 \to \mathcal{L}_2} = \|\tilde{M}\|_{\ell_2 \to \ell_2} = \|\check{M}\|_\infty$$

where $\mathbf{M} = \mathcal{F}_l(\mathbf{G}, \mathbf{HK}_d\mathbf{S})$, the nominal plant and controller arrangement.

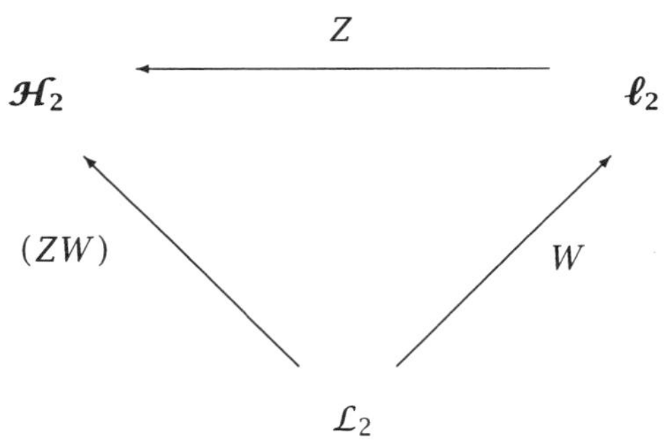

Figure 3.3: Signal Spaces

Recall that the multiplication operator $\Theta_{\check{M}}$ is the image of $\tilde{M}$ under the z-transform. In the sequel we will move between these three representations for an h-periodic operator on $\mathcal{L}_2$. Figure 3.3 shows the connections between the domains of these representations. As we have done so far, we will continue to adhere to the following convention: given an operator $\mathbf{T}$ on $\mathcal{L}_2$, we use the notation $\tilde{T}$ to denote its image as an operator on ℓ_2, and if it is h-periodic, $\check{T}$ to signify its transfer function in $\mathcal{H}_\infty$.

The initial feedback system of Figure 3.1 has now been transformed to the equivalent representation in Figure 3.4. The key feature of this transformed system is that $\tilde{M}$ is LTI on the space of signals ℓ_2.

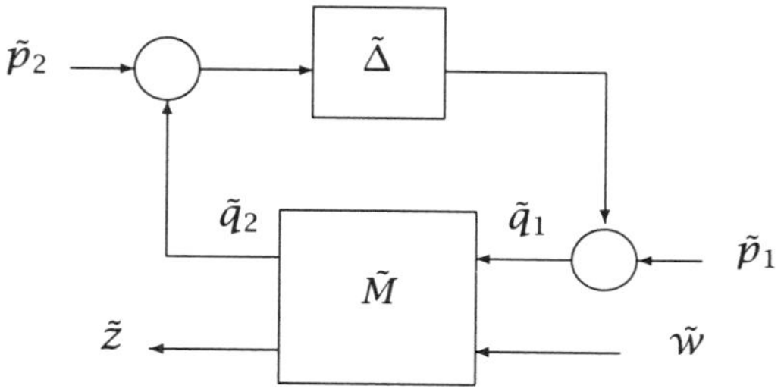

Figure 3.4: LTI System

Conformally partition the operator $\tilde{M}$ with respect to the inputs and outputs of the new block diagram:

$$\tilde{M} =: \begin{bmatrix} \tilde{M}^{11} & \tilde{M}^{12} \\ \tilde{M}^{21} & \tilde{M}^{22} \end{bmatrix}.$$

Referring to this partition we have the following robust stability result.

Lemma 3.4 *Suppose that $\mathfrak{X}$ is a subspace of $\mathfrak{L}(\mathcal{L}_2)$ and $r > 0$. Then the system in Figure 3.1 is robustly stabilized to $r\mathcal{U}\mathfrak{X}$ if and only if for each $\Delta \in r\mathcal{U}\mathfrak{X}$ the map $(I - \tilde{M}^{11}\tilde{\Delta})^{-1}$ exists in $\mathfrak{L}(\ell_2)$.*

This lemma states that robust stability of the sampled-data system is equivalent to the existence and boundedness of one of the maps in Figure 3.4 for all perturbations in the uncertainty set.

Proof Set $\mathbf{M} = W^{-1}\tilde{M}W = \mathcal{F}_l(\mathbf{G}, \mathbf{HK}_d\mathbf{S})$ the nominal system in Figure 3.1. (Only if): The map $p_2 \mapsto q_2$ is bounded by assumption and is given by $(I - \mathbf{M}^{11}\boldsymbol{\Delta})^{-1} - I$. Therefore, $W(I - \mathbf{M}^{11}\boldsymbol{\Delta})^{-1}W^{-1}$ is also bounded, and is equal to $(I - \tilde{M}^{11}\tilde{\Delta})^{-1}$.

(If): Similarly, since $(I - \tilde{M}^{11}\tilde{\Delta})^{-1}$ is bounded for each $\Delta \in r\mathcal{U}\mathfrak{X}$ the map $(I - \mathbf{M}^{11}\Delta)^{-1}$ is bounded on $\mathcal{L}_2$. Now, $I + \Delta(I - \mathbf{M}^{11}\Delta)^{-1}\mathbf{M}^{11} = (I - \Delta\mathbf{M}^{11})^{-1}$ and is therefore also bounded.

From the last two statements we can conclude stability of the system in Figure 3.1 because all the maps $w, p_1, p_2 \mapsto z, q_1, q_2$ are given by products and sums of the operators Δ, $(I - \Delta\mathbf{M}^{11})^{-1}$, $(I - \mathbf{M}^{11}\Delta)^{-1}$, and the components of $\mathbf{M}$. ■

Stated below is a well-known test for robust stability — the so-called *small gain theorem*. This result has played a central role in robust control research, and was first proved by Zames [74] and Sandberg [57].

Theorem 3.5 *Suppose that* $\mathfrak{X} \subset \mathfrak{L}(\mathcal{L}_2)$ *and* $r > 0$. *If* $r\|\mathbf{M}^{11}\|_{\mathcal{L}_2 \to \mathcal{L}_2} \leq 1$, *then the system in Figure 3.1 has robust stability to* $r\mathcal{U}\mathfrak{X}$.

The result provides a sufficient robustness condition for any subspace $\mathfrak{X}$ of $\mathfrak{L}(\mathcal{L}_2)$. The necessity of this condition depends on the characteristics of *both* the system $\mathbf{M}$ and the uncertainty set $\mathfrak{X}$. As stated above with $\mathfrak{X} = \mathfrak{L}(\mathcal{L}_2)$ the result can be strengthened to being both necessary and sufficient for any nominal system $\mathbf{M}$ in $\mathfrak{L}(\mathcal{L}_2)$. However for the structured uncertainty set $\mathfrak{X}_s$ defined in (3.2), and its subsets, the *small gain condition* $r\|\mathbf{M}^{11}\|_{\mathcal{L}_2 \to \mathcal{L}_2} \leq 1$ is not in general necessary for robust stability, and is typically a conservative test. The sequel will be concerned with these more complex uncertainty structures.

Proof Choose any $\Delta \in \mathcal{U}\mathfrak{X}$. By Lemma 3.4 it is sufficient to show that $I - \mathbf{M}^{11}\Delta$ is invertible. Since $\|\mathbf{M}^{11}\|_{\mathcal{L}_2 \to \mathcal{L}_2} \leq r^{-1}$ we have

$$\|\mathbf{M}^{11}\Delta\|_{\mathcal{L}_2 \to \mathcal{L}_2} \leq \|\mathbf{M}^{11}\|_{\mathcal{L}_2 \to \mathcal{L}_2} \cdot \|\Delta\|_{\mathcal{L}_2 \to \mathcal{L}_2} < 1.$$

Thus by Proposition 2.1 we have that $\mathrm{rad}(\mathbf{M}^{11}\Delta) < 1$, which ensures that $1 \notin \mathrm{spec}(\mathbf{M}^{11}\Delta)$. ■

We have provided a general setting for studying the sampled-data robustness issues of subsequent chapters. In the next chapter we consider robust stabilization when our uncertainty model consists of LTI perturbations.

3.1 Summary

After formally defining sampled-data systems we have introduced a general class of structured uncertainty problems. Namely the problems of determining exact conditions for robust stability and performance, defined in Definitions 3.2 and 3.3, to particular causal subsets of $\mathfrak{X}_s$. Developing approaches and answers to these problems is addressed in subsequent chapters.

In this book the signal space of interest is $\mathcal{L}_2$. However the related work of Sivashankar and Khargonekar [59], and Khammash [37], treats sampled-data robust performance with $\mathcal{L}_\infty$ signals where the uncertainty is structured and time-varying.

Considerable research has been devoted to studying sampled-data systems without uncertainty, and sampled-data formulations and solutions exist to all the standard control synthesis problems. Minimizing the $\mathcal{L}_2$ induced norm, the so-called sampled-data $\mathcal{H}_\infty$ problem, is solved using a variety of techniques in Hara and Kabamba [32], Bamieh and Pearson [7], Toivonen [66], Basar and Bernhard [9], and Sun, Nagpal and Khargonekar [62]. The sampled-data $\mathcal{H}_2$ problem has been considered by Chen and Francis [18], Bamieh and Pearson [6], and Sivashankar and Khargonekar [60]. Sampled-data $\mathcal{L}_1$ synthesis (minimizing the $\mathcal{L}_\infty$ induced norm) is studied in Dullerud and Francis [24], and Bamieh, Dahleh and Pearson [5].

Multirate sampled-data synthesis has been considered by Chen and Qui in [15] and [51], for $\mathcal{H}_2$ and $\mathcal{H}_\infty$ performance respectively. The asynchronous multirate $\mathcal{H}_\infty$ sampled-data problem is treated in Lall [38].

Model validation has also been considered in the sampled-data context of this chapter. Validation of continuous time models with uncertainty, given finite sampled-data records, is investigated in Ragnan and Poolla [52], and Smith and Dullerud [61].

Chapter 4

Analysis of LTI Uncertainty

Our intent now is to construct a robust stability condition which characterizes system robustness to stable LTI perturbations. Furthermore, we continue to develop a general framework in which to consider issues of robustness for sampled-data systems.

The class of all causal LTI perturbations can be represented by the transfer function space $\mathcal{H}_\infty$. Instead of using the entire space of LTI perturbations, we will select those that have transfer functions $\widehat{\Delta}$ in the half plane algebra $\mathcal{A}_{\mathbb{R}}$. By Proposition 2.12 this perturbation class is the closure of the FDLTI operators. It therefore forms a natural perturbation set for the FDLTI plant, and also greatly expedites our analysis because of the better boundary properties of the functions in $\mathcal{A}_{\mathbb{R}}$.

The robust stabilization problem we consider is based on the arrangement shown in Figure 4.1. By Lemma 3.4 robust stability of this configuration is immediately captured by the setup in Figure 3.4 by setting $\tilde{M}$ to $\tilde{M}^{11}$.

The perturbation in Figure 4.1 is assumed to be a causal LTI operator $\boldsymbol{\Delta}$ on $\mathcal{L}_2$, and is in an LTI subset of the spatial uncertainty set $\mathfrak{X}_s$ defined in Chapter 3. We define this LTI subset using the subset of matrices in $\mathbb{C}^{m\times m}$ given by

$$\mathcal{X} := \{\mathrm{diag}(\underline{\Delta}_1, \dots, \underline{\Delta}_d) \in \mathbb{C}^{m\times m} : \; \underline{\Delta}_k \in \mathbb{C}^{m_k\times m_k}\}, \tag{4.1}$$

for a fixed set of d positive integers m_k, which must be compatible with

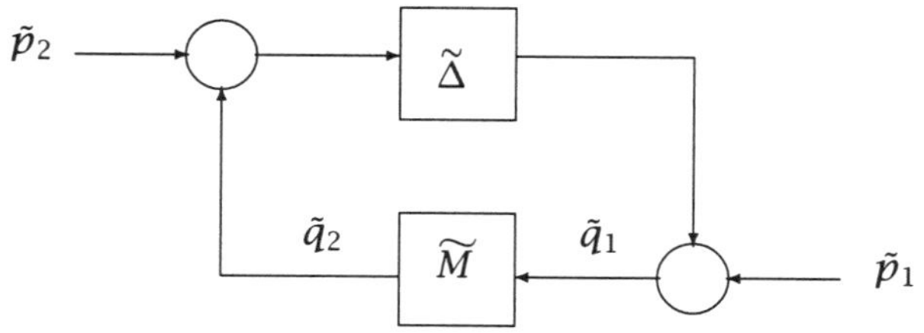

Figure 4.1: Robust Stabilization Model

m. Note: all the subsequent results also hold with the set $\mathcal{X}$ defined more generally as in (2.10). Next, from the set of spatially structured matrices $\mathcal{X}$, we define a subset of $\mathcal{A}_{\mathbb{R}}$ consisting of functions that map $\bar{\mathbb{C}}_+$ to $\mathcal{X}$:

$$\mathcal{A}_{\mathbb{R}}^{\mathcal{X}} := \{\hat{\Delta} \in \mathcal{A}_{\mathbb{R}} : \hat{\Delta}(s_0) \in \mathcal{X} \text{ for all } s_0 \in \bar{\mathbb{C}}_+\}.$$

We denote the space of linear operators on $\mathcal{L}_2$ that correspond to these transfer functions by

$$\mathfrak{X}_{LTI}$$

and note that its spatial structure makes it a subspace of $\mathfrak{X}_s$.

Precisely stated, the main aim of this chapter is to develop a necessary and sufficient condition for robust stabilization, as defined in Definition 3.2, of the system in Figure 4.1 to the perturbation class $\mathfrak{X}_{LTI}$.

In constructing the above robustness condition we will change the domain in which we work: above we have concisely represented our LTI perturbations by the algebra $\mathcal{A}_{\mathbb{R}}^{\mathcal{X}}$; although this is a convenient way to define the uncertainty, it is not an advantageous domain in which to consider sampled-data robustness. In the sequel we find that a better viewpoint is gained by posing robustness in the operator-valued frequency domain defined by $\boldsymbol{\mathcal{A}}$. This allows us to see the structure of LTI robustness, and in this chapter we find it efficient to change domains again and formulate our final robustness criterion in terms of a function taking values in $\mathfrak{L}(\ell_2)$.

We remark that all the results of the next three sections hold for a more general nominal plant than the feedback configuration of Figure 3.1 where $\mathbf{M} = \mathcal{F}_l(\mathbf{G}, \mathbf{HK}_d\mathbf{S})$. The only property required of $\mathbf{M}$ is that it has a transfer function $\check{M}$ that is in $\mathcal{A}$, and that at each z_0 in $\bar{\mathbb{D}}$ the operator $\check{M}(z_0)$ is compact. Thus the results developed here are applicable to a wider class of h-periodic systems.

The chapter is partitioned into four results sections, each of which addresses a particular step in the analysis: first we convert our entire problem to one posed in the frequency domain space $\mathcal{A}$; then in Section 4.2 we refine this frequency domain test, and characterize the space $\mathcal{A}_{\mathbb{R}}$ as a subspace of $\mathcal{A}$. The conditions we seek are constructed in Section 4.3. The section of final results develops some properties of the *frequency response* function which will defined during the chapter.

4.1 Converting to Frequency Domain

Our main task now is to pose the robust stabilization problem in the frequency domain. To start, we show that the operators in $\mathfrak{L}_{\mathcal{A}_{\mathbb{R}}}$, arising from the transfer functions $\hat{\Delta}$ in $\mathcal{A}_{\mathbb{R}}$, can be represented by functions in $\mathcal{A}$; and therefore form a real subspace of $\mathcal{A}$.

Lemma 4.1 *If $\Delta \in \mathfrak{L}_{\mathcal{A}_{\mathbb{R}}}$, then there exists a unique function $\check{\Delta} \in \mathcal{A}$ so that $Z^{-1}\Theta_{\check{\Delta}}Z = W\Delta W^{-1}$. Furthermore, $\|\Delta\|_{\mathcal{L}_2 \to \mathcal{L}_2} = \|\check{\Delta}\|_\infty$.*

Proof By Proposition 2.12 we know that there exist stable FDLTI operators $\Delta_k \in \mathfrak{L}_{\mathcal{A}_{\mathbb{R}}}$ so that

$$\lim_{k \to \infty} \|\Delta - \Delta_k\|_{\mathcal{L}_2 \to \mathcal{L}_2} = 0.$$

Since $\|\Delta - \Delta_k\|_{\mathcal{L}_2 \to \mathcal{L}_2} = \|W(\Delta - \Delta_k)W^{-1}\|_{\ell_2 \to \ell_2}$, we have that $\|W(\Delta - \Delta_k)W^{-1}\|_{\ell_2 \to \ell_2} \to 0$ as $k \to \infty$.

From (2.5) and (2.8), in Chapter 2 we know that there exist $\check{\Delta}_k \in \mathcal{A}$ so that $Z^{-1}\Theta_{\check{\Delta}_k}Z = W\Delta_k W^{-1}$. Now,

$$\|\check{\Delta}_k - \check{\Delta}_l\|_\infty = \|W(\Delta_k - \Delta_l)W^{-1}\|_{\ell_2 \to \ell_2},$$

and therefore $\check{\Delta}_k$ is a Cauchy sequence in $\mathcal{A}$. By the completeness of $\mathcal{A}$, Proposition 2.11, there exists a unique $\check{\Delta} \in \mathcal{A}$ so that $\|\check{\Delta} - \check{\Delta}_k\|_\infty \to 0$

as $k \to \infty$. Therefore, $Z^{-1}\Theta_{\check{\Delta}}Z = W\Delta W^{-1}$. The equivalence of the norms follows directly from the fact that Z and W are isometric maps.■

The lemma tells us that for each operator $\Delta \in \mathfrak{L}_{\mathcal{A}_{\mathbb{R}}}$ we can find its isometric isomorphic image $\check{\Delta} \in \mathcal{A}$. The relationship between the various spaces is summarized by the commutative diagram in Figure 4.2.

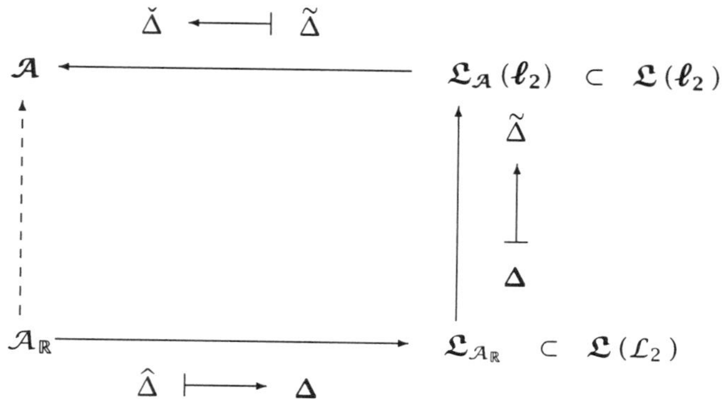

Figure 4.2: Relationships Between the Spaces

In the diagram $\mathfrak{L}_{\mathcal{A}}(\ell_2)$ signifies the operators on ℓ_2 which have transfer functions in $\mathcal{A}$; they are isomorphic to the space of operators $\mathfrak{L}_{\mathcal{A}}$ on $\mathcal{L}_2$. We will use

$$\mathcal{A}_{\mathcal{A}_{\mathbb{R}}}$$

to denote the image of $\mathcal{A}_{\mathbb{R}}$ in the space $\mathcal{A}$.

To explicitly illustrate the mappings between these spaces we take a concrete example: consider the function

$$\hat{T}(s) = \frac{e^{-sh/2}}{s+1},$$

which is in $\mathcal{A}_{\mathbb{R}}$. In $\mathfrak{L}_{\mathcal{A}_{\mathbb{R}}}$ it corresponds to the operator

$$\mathbf{T} = \mathbf{D}_{\frac{h}{2}}\mathbf{F},$$

where $\mathbf{D}_{\frac{h}{2}}$ is a delay operator with delay $\frac{h}{2}$, and $\mathbf{F}$ is the FDLTI operator with realization $A = -1$, $B = 1$, $C = 1$, and $D = 0$. We can obtain a representation for $\tilde{D}_{\frac{h}{2}}$, defined by $\tilde{D}_{\frac{h}{2}} := W\mathbf{D}_{\frac{h}{2}}W^{-1}$, in $\mathfrak{L}_{\mathcal{A}}$. Define the left and right shift operators on $\mathcal{K}_2$, for $\psi \in \mathcal{K}_2$:

$$(R_{\frac{h}{2}}\psi)(\tau) := \begin{cases} 0, & \tau \in [0, \frac{h}{2}) \\ \psi(\tau - \frac{h}{2}), & \tau \in [\frac{h}{2}, h) \end{cases}$$

$$(L_{\frac{h}{2}}\psi)(\tau) := \begin{cases} \psi(\tau + \frac{h}{2}), & \tau \in [0, \frac{h}{2}) \\ 0, & \tau \in [\frac{h}{2}, h) \end{cases}$$

It is routine to see that $\tilde{D}_{\frac{h}{2}}$ can be described, for $\tilde{u} \in \boldsymbol{\ell}_2$, by

$$(\tilde{D}_{\frac{h}{2}}\tilde{u})[k] := \begin{cases} R_{\frac{h}{2}}\tilde{u}[k] + L_{\frac{h}{2}}\tilde{u}[k-1], & k > 0 \\ R_{\frac{h}{2}}\tilde{u}[0], & k = 0 \end{cases}$$

Now, by setting $\tilde{y} = \tilde{D}_{\frac{h}{2}}\tilde{u}$ and taking z-transforms, it is easy to show that $\check{y}(z) = R_{\frac{h}{2}}\check{u}(z) + zL_{\frac{h}{2}}\check{u}(z)$. Hence, $\check{D}_{\frac{h}{2}}(z) = R_{\frac{h}{2}} + zL_{\frac{h}{2}}$. We already know how $\mathbf{F}$ maps to $\check{F}$ in $\mathcal{A}$, by (2.5) and (2.8), therefore we have

$$\check{T}(z) = \check{D}_{\frac{h}{2}}(z)\check{F}(z).$$

We now take our stability condition which is stated in Definition 3.2 purely in terms of operators on $\mathfrak{L}(\mathcal{L}_2)$, and convert it to an equivalent frequency domain condition on functions in $\mathcal{A}$. Recall that $\check{M}$ is the sampled-data transfer function defined in (3.5), and that $\mathfrak{L}_{\mathcal{A}}$ is the space of operators on $\mathcal{L}_2$ with transfer functions in $\mathcal{A}$. We have:

Lemma 4.2 *Given the previous definitions and suppose $\Delta \in \mathfrak{L}_{\mathcal{A}}$. Then the map $(I - \widetilde{M}\widetilde{\Delta})^{-1}$ exists and is bounded if and only if for each z in the closed unit disk $\bar{\mathbb{D}}$, the operator $I - \check{M}(z)\check{\Delta}(z)$ is nonsingular.*

Proof (Only if): By assumption $(I - \widetilde{M}\widetilde{\Delta})^{-1}$ exists and is bounded. Therefore $(\Theta_{I-\check{M}\check{\Delta}})^{-1}$ exists and is bounded on $\mathcal{H}_2$, because of the relationship $Z^{-1}\Theta_{I-\check{M}\check{\Delta}}Z = (I - \widetilde{M}\widetilde{\Delta})$; hence $(I - \check{M}\check{\Delta})^{-1}$ is in $\mathcal{H}_\infty$ by Proposition 2.10 since $(I - \widetilde{M}\widetilde{\Delta})^{-1}$ is LTI.

Suppose there exists a complex number $z' \in \partial\mathbb{D}$ so that $I - \check{M}(z')\check{\Delta}(z')$ is not invertible. Then by continuity of $I - \check{M}(z)\check{\Delta}(z)$ on $\bar{\mathbb{D}}$, we have that

$\|(I - \check{M}(z)\check{\Delta}(z))^{-1}\| \to \infty$ as $z \to z'$, see for example [3, pp. 41], contradicting that $(I - \check{M}\check{\Delta})^{-1} \in \mathcal{H}_\infty$. Therefore no such $z' \in \partial\mathbb{D}$ exists.

(If): It is sufficient to show that $(I - \check{M}\check{\Delta})^{-1} \in \mathcal{A}$.

We begin with continuity and boundedness on $\bar{\mathbb{D}}$: these both follow from the inequality

$$\|(I - \check{M}\check{\Delta})^{-1}(z) - (I - \check{M}\check{\Delta})^{-1}(z_0)\| \le \frac{\|(I - \check{M}\check{\Delta})^{-1}(z_0)\|^2 \, \|(\check{M}\check{\Delta})(z) - (\check{M}\check{\Delta})(z_0)\|}{1 - \|(I - \check{M}\check{\Delta})^{-1}(z_0)\|^{-1} \, \|(\check{M}\check{\Delta})(z) - (\check{M}\check{\Delta})(z_0)\|},$$

which holds when $\|(\check{M}\check{\Delta})(z) - (\check{M}\check{\Delta})(z_0)\|$ is less than $\|(I - \check{M}\check{\Delta})^{-1}(z_0)\|$. See for example [11, p. 170] for this standard inverse inequality.

For analyticity, see [34, pp. 96], $(I - (\check{M}\check{\Delta})(z))^{-1}$ is analytic at each point $z \in \mathbb{D}$, because $(I - (\check{M}\check{\Delta})(z))^{-1}$ exists and $I - (\check{M}\check{\Delta})(z)$ is analytic at every $z \in \mathbb{D}$. ■

Our robustness problem can now be reformulated as a frequency domain test in the space $\mathcal{A}$ of operator-valued transfer functions:

Corollary 4.3 *Given the system in Figure 4.1 and a subset $\mathfrak{X}$ of $\mathfrak{L}_{\mathcal{A}}$. The system is robustly stabilized to $\mathcal{U}\mathfrak{X}$ if and only if for every $\check{\Delta} \in \mathcal{U}\mathcal{A}_{\mathfrak{X}}$ the operator $I - \check{M}(z)\check{\Delta}(z)$ is nonsingular for each $z \in \bar{\mathbb{D}}$, where $\mathcal{A}_{\mathfrak{X}}$ is the representation of $\mathfrak{X}$ in $\mathcal{A}$.*

The claim follows directly from Lemmas 3.4 and 4.2.

From now on we will concentrate exclusively on the frequency domain. In the next subsection we obtain a description of perturbations which destabilize the feedback system.

4.2 Destabilizing Perturbations

In the previous subsection we showed that the functions $\hat{\Delta} \in \mathcal{A}_{\mathbb{R}}$ could be represented by functions in $\mathcal{A}$. We then converted our robust stabilization problem to an equivalent test on $\mathcal{A}$, which was defined by function evaluations on $\bar{\mathbb{D}}$. Here the goals are to examine the transfer function properties on the unit circle $\partial\mathbb{D}$, and to obtain an explicit representation for the functions $\hat{\Delta} \in \mathcal{A}_{\mathbb{R}}$ when they are mapped to $\check{\Delta}$ in $\mathcal{A}$, as shown by the broken line in Figure 4.2.

The first lemma we require provides a maximum modulus type result for the spectral radius functions of elements in $\mathcal{A}$.

Lemma 4.4 *Suppose that $\check{F} \in \mathcal{A}$. Then*

$$\max_{z\in\partial\mathbb{D}} \mathrm{rad}(\check{F}(z)) = \max_{z\in\bar{\mathbb{D}}} \mathrm{rad}(\check{F}(z)).$$

This lemma states that the maximum spectral radius of such a function is achieved on the boundary of the disc.

Proof We begin by noting that $\mathrm{rad}(\check{F}(z))$ is an upper semicontinuous scalar function on $\bar{\mathbb{D}}$. This is because $\mathrm{rad}(\cdot)$ is upper semicontinuous on $\mathfrak{L}(\mathcal{K}_2)$, and $\check{F}(z)$ is continuous on $\bar{\mathbb{D}}$. Therefore, the maxima above are well-defined on the compact sets $\bar{\mathbb{D}}$ and $\partial\mathbb{D}$.

By the result stated in Proposition 2.7, $\mathrm{rad}(\check{F}(z))$ is a subharmonic function on $\mathbb{D}$ because $\check{F}(z)$ is analytic there. Hence, $\mathrm{rad}(\check{F}(z))$ satisfies the maximum principle of Proposition 2.4. There are two possibilities: $\mathrm{rad}(\check{F}(z))$ has no maximum value on $\mathbb{D}$, or $\mathrm{rad}(\check{F}(z))$ is constant on $\mathbb{D}$.

In the first case we have immediately that the maximum of $\mathrm{rad}(\check{F}(z))$ on $\bar{\mathbb{D}}$, since it exists, must be attained on $\partial\mathbb{D}$.

Suppose that $\mathrm{rad}(\check{F}(z))$ is constant on $\mathbb{D}$. Then by upper semicontinuity of $\mathrm{rad}(\check{F}(z))$ on $\bar{\mathbb{D}}$, for any $z' \in \mathbb{D}$ and $z'' \in \partial\mathbb{D}$ we have that $\mathrm{rad}(\check{F}(z'')) \geq \mathrm{rad}(\check{F}(z'))$. ■

With this maximum modulus result we can prove the next lemma which provides a necessary and sufficient test for robustness to a subspace in terms of function evaluations on the unit circle.

Lemma 4.5 *Suppose $\mathcal{X}$ is a complex subspace of $\mathcal{A}$. Then the spectral radius function $\mathrm{rad}((\check{M}\check{\Delta})(z)) < 1$ on $\partial\mathbb{D}$ for all $\check{\Delta} \in \mathcal{U}\mathcal{X}$, if and only if, for all $\check{\Delta} \in \mathcal{U}\mathcal{X}$ the map $I - (\check{M}\check{\Delta})(z)$ is nonsingular on $\bar{\mathbb{D}}$.*

Proof (Only if): By the contrapositive: suppose there exists $z' \in \bar{\mathbb{D}}$ and $\check{\Delta} \in \boldsymbol{X}$ so that $I - (\check{M}\check{\Delta})(z')$ is not invertible, then $\text{rad}(\check{M}\check{\Delta})(z') \geq 1$. By Lemma 4.4 there exists a $z'' \in \partial\mathbb{D}$ so that $\text{rad}((\check{M}\check{\Delta})(z'')) \geq 1$.
(If): Suppose that there exists $\check{\Delta} \in \boldsymbol{X}$ so that $1 \leq \text{rad}((\check{M}\check{\Delta})(z'))$ at some $z' \in \bar{\mathbb{D}}$. Let $\lambda \in \text{spec}((\check{M}\check{\Delta})(z))$ be such that $|\lambda| = \text{rad}((\check{M}\check{\Delta})(z))$. Then $\check{\Delta}_0 := \lambda^{-1}\check{\Delta} \in \mathcal{U}\boldsymbol{X}$ since $\boldsymbol{X}$ is a subspace, and $I - (\check{M}\check{\Delta}_0)(z')$ is singular. ■

Note however that our perturbation class $\boldsymbol{\mathcal{A}}_{\mathcal{A}_{\mathbb{R}}}$ is not a complex subspace of $\boldsymbol{\mathcal{A}}$; it is a real subspace. Hence, the above result only provides a sufficient condition for robust stability. Later we will show that this sufficient condition is also a necessary one.

By Lemma 4.1 we know that $\mathfrak{L}_{\mathcal{A}_{\mathbb{R}}}$ can be represented as a subspace of $\boldsymbol{\mathcal{A}}$; however we still do not know the form of this mapping. We now seek an explicit representation of our perturbations $\Delta \in \mathfrak{L}_{\mathcal{A}_{\mathbb{R}}}$, when they are transformed, as in Lemma 4.1, to $\check{\Delta} \in \boldsymbol{\mathcal{A}}$. Because Lemma 4.5 states that a boundary check of the transfer functions is a sufficient condition for robust stabilization, we concentrate on capturing the values of $\check{\Delta}$ on the boundary of the disc. This will eventually lead to the conclusion that checking the boundary properties of these transfer functions is both necessary and sufficient for robust stabilization.

Begin by defining the sequences $\{\phi_k\}_{k=0}^{\infty}$ and $\{\psi_k\}_{k=0}^{\infty}$ of scalar valued orthonormal functions in $\mathcal{K}_2$:

$$\phi_k(t) := h^{-\frac{1}{2}} e^{jh^{-1}2\pi\nu_k t}, \qquad t \in [0, h) \tag{4.2}$$

$$\psi_k(t) := h^{-\frac{1}{2}} e^{jh^{-1}(2\pi\nu_k - \omega_0)t}, \qquad t \in [0, h), \tag{4.3}$$

where ω_0 is any fixed number in the interval $(-\pi, \pi]$, and ν_k is the sequence
$\{0, 1, -1, 2, -2, \ldots\}$. It is a standard fact that the functions $\{\phi_k\}_{k=0}^{\infty}$ form a complete orthonormal basis for $\mathcal{K}_2$. Pertaining to the functions $\{\psi_k\}_{k=0}^{\infty}$ we have a similar result:

Proposition 4.6 *The set of functions $\{\psi_k\}_{k=0}^{\infty}$, defined in (4.3), forms a complete orthonormal basis for $\mathcal{K}_2$.*

Proof With a slight abuse of notation, suppose $\psi(t) \in \mathcal{K}_2$ is a scalar valued function. Then $e^{-jh^{-1}\omega_0 t}\psi(t) \in \mathcal{K}_2$. Hence, the function has an expansion in terms of $\phi_k(t)$ defined in (4.2): $e^{-jh^{-1}\omega_0 t}\psi(t) =: \sum_{k=0}^{\infty} a_k\phi_k(t)$ with the coefficients $a_k \in \mathbb{C}$. This implies that $\psi(t) = \sum_{k=0}^{\infty} a_k\psi_k(t)$ because $e^{jh^{-1}\omega_0 t}\phi_k(t) = \psi_k(t)$.■

This basis can be used to find Fourier expansions for vector valued functions in $\mathcal{K}_2^m$: suppose that $\psi \in \mathcal{K}_2^m$; then there exist $a_k \in \mathbb{C}^m$ so that

$$\psi = \sum_{k=0}^{\infty} a_k\psi_k.$$

We will therefore say that $\{\psi_k\}_{k=0}^{\infty}$ forms a complete orthonormal basis for $\mathcal{K}_2$ regardless of the spatial dimension, where it is understood that the coefficients in any expansion are appropriately dimensioned complex vectors. Note that the basis depends on the parameter ω_0, although this is not included in our notation.

The following lemma provides an exact connection between the operator-valued transfer function $\check{\Delta}$ and the transfer function $\hat{\Delta}$.

Lemma 4.7 *Suppose that $\psi = a\psi_k$ for a fixed integer $k \geq 0$, $a \in \mathbb{C}^m$, and $\omega_0 \in (-\pi, \pi]$. If $\mathbf{\Delta} \in \mathfrak{L}_{\mathcal{A}_\mathbb{R}}$, then $\check{\Delta}(e^{j\omega_0})\psi = \hat{\Delta}(j\frac{2\pi\nu_k-\omega_0}{h})\psi$.*

The results states that ψ is an eigenvector of the operator $\check{\Delta}(e^{j\omega_0})$, evaluated at any frequency ω_0.

Proof By Proposition 2.12 we can choose a sequence $\{\Delta_n\}_{n=0}^{\infty}$ of stable FDLTI operators so that $\Delta_n \to \mathbf{\Delta}$ as $n \to \infty$. By the triangle inequality

$$\begin{aligned}&\|(\check{\Delta}(e^{j\omega_0}) - \hat{\Delta}(j\tfrac{2\pi\nu_k-\omega_0}{h})I)\psi + (\hat{\Delta}_n(j\tfrac{2\pi\nu_k-\omega_0}{h})I - \check{\Delta}_n(e^{j\omega_0}))\psi\|_2 \\ &\leq \|(\check{\Delta}(e^{j\omega_0}) - \check{\Delta}_n(e^{j\omega_0}))\psi\|_2 + \|(\hat{\Delta}_n(j\tfrac{2\pi\nu_k-\omega_0}{h}) - \hat{\Delta}(j\tfrac{2\pi\nu_k-\omega_0}{h}))\psi\|_2.\end{aligned}$$

Now, for any stable FDLTI map with minimal realization (A, B, C, D) it is routine to verify, from the formulae given in (2.6) and (2.8), that

$$(\check{C}e^{j\omega_0}(I - e^{j\underline{\omega}_0}\check{A})^{-1}\check{B} + \check{D})\psi = (C(Ih^{-1}(2\pi\nu_k - \omega_0) - A)^{-1}B + D)\psi.$$

Therefore, $(\widehat{\Delta}_n(j\frac{2\pi\nu_k-\omega_0}{h})I - \check{\Delta}_n(e^{j\omega_0}))\psi = 0$ for all n. Our inequality becomes therefore

$$
\begin{aligned}
&\|(\check{\Delta}(e^{j\omega_0}) - \widehat{\Delta}(j\tfrac{2\pi\nu_k-\omega_0}{h})I)\psi\|_2 \\
&\leq \|(\check{\Delta}(e^{j\omega_0}) - \check{\Delta}_n(e^{j\omega_0}))\psi\|_2 + \|(\widehat{\Delta}_n(j\tfrac{2\pi\nu_k-\omega_0}{h}) - \widehat{\Delta}(j\tfrac{2\pi\nu_k-\omega_0}{h}))\psi\|_2
\end{aligned}
$$

The RHS above tends to zero as $k \to \infty$, because $\|\check{\Delta} - \check{\Delta}_n\|_\infty = \|\widehat{\Delta}_n - \widehat{\Delta}\|_\infty = \|\Delta - \Delta_n\|_{\mathcal{L}_2\to\mathcal{L}_2}$, and by definition $\|\Delta - \Delta_n\|_{\mathcal{L}_2\to\mathcal{L}_2} \to 0$. Therefore, the LHS =0. ■

The above lemma therefore gives us an exact characterization of the boundary values of $\check{\Delta}$ in terms of the values of $\widehat{\Delta}$, since the set $\{\psi\}_{k=0}^{\infty}$ spans all of $\mathcal{K}_2$. We remark that it is possible to extend the definition of the basis into the disc so that Lemma 4.7 holds. Unfortunately, this basis, when extended inside the disc, is no longer orthonormal.

We have the following boundary description of the $\mathcal{A}_{\mathcal{A}_{\mathbb{R}}}$ functions:

Corollary 4.8 *Given $\omega_0 \in (-\pi, \pi]$ and the infinite sequence $\theta_k := \frac{2\pi\nu_k-\omega_0}{h}$. If $\Delta \in \mathfrak{L}_{\mathcal{A}_{\mathbb{R}}}$, then $\check{\Delta}(e^{j\omega_0}) = diag\{\widehat{\Delta}(j\theta_k)\}_{k=0}^{\infty}$ with respect to the basis functions $\{\psi_k\}_{k=0}^{+\infty}$.*

This result follows directly from the previous lemma and proposition. It states that the operator $\check{\Delta}(e^{j\omega_0})$ is an infinite block diagonal matrix with respect to the basis $\{\psi_k\}_{k=0}^{+\infty}$. Also, the diagonal blocks are exactly the matrices $\widehat{\Delta}(j\frac{2\pi k-\omega_0}{h})$ from the related continuous-time transfer function in $\mathcal{A}_{\mathbb{R}}$. The commutative diagram in Figure 4.2 shows this mapping by the dashed line.

Since $\check{\Delta}(e^{j\omega_0})$ is completely determined by the values of $\widehat{\Delta}(j\omega)$, we investigate the allowable behavior of $\widehat{\Delta}(j\omega)$. We prove two technical lemmas that help us to show the interpolation result required.

Lemma 4.9 *Suppose $\omega_0 \in (-\pi, \pi)$, $\omega_0 \neq 0$, and θ_k is the infinite sequence $\{\frac{2\pi\nu_k-\omega_0}{h}\}_{k=0}^{\infty}$. Given a finite set of distinct non-negative integers $K = \{k_0, \ldots, k_n\}$, there exists $f \in \mathcal{A}_{\mathbb{R}}$, with $\|f\|_\infty = 1$, satisfying*

$$
f(j\theta_k) = \begin{cases} 1 & k \in K \\ 0 & k \notin K \end{cases}.
$$

It is clear from the proof below that this result can be extended to any sequence of points, on the imaginary axis, providing the interpolation constraints do not conflict with the fact that f is in $\mathcal{A}_{\mathbb{R}}$ and thus continuous. Our constructive proof is based on a more general existence proof given in [55], where the set of interpolation points on the imaginary axis is only assumed to have Lebesgue measure zero.

Proof We construct f from two functions:

$$\begin{aligned} p(s) &:= \sum_{l=0}^{n} \sqrt{\frac{s}{s^2+\theta_{k_l}^2}+1} \\ q(s) &:= \sum_{\substack{k=0 \\ k\notin K}}^{\infty} \left(\frac{1}{2}\right)^k \sqrt{\frac{s}{s^2+\theta_k^2}+1}. \end{aligned}$$

It is straightforward to verify that $p(s)$ and $q(s)$ are analytic in $\mathbb{C}_+$, and continuous on $\bar{\mathbb{C}}_+$ except at the points $j\theta_k$ where they are infinite. They also satisfy $p(s^*) = p(s)^*$ and $q(s^*) = q(s)^*$. Furthermore, it is routine to see that their phase functions only take values in the interval $(-\frac{\pi}{4}, \frac{\pi}{4})$, and therefore the real part of q/p is always non negative. Hence,

$$f(s) := \frac{1}{1+q/p}$$

satisfies the interpolation criteria, has norm of one, and is in $\mathcal{A}_{\mathbb{R}}$ since the interpolation constraints are consistent with f being continuous. ■

We now generalize the above result to interpolation of matrices in the set $\mathcal{X}$ which was defined in (4.1).

Lemma 4.10 *Suppose $\omega_0 \in (-\pi, \pi)$, $\omega_0 \neq 0$, and θ_k is the infinite sequence $\{\frac{2\pi v_k - \omega_0}{h}\}_{k=0}^{\infty}$. Given a complex matrix Δ in the set $\mathcal{X}$ with $\bar{\sigma}(\Delta) < 1$, a real number γ in the interval $(0, 1)$, and a finite set of distinct non-negative integers $K = \{k_0, \ldots, k_n\}$, there exists $\hat{\Delta} \in \mathcal{UA}_{\mathbb{R}}^{\mathcal{X}}$ so that*

$$\hat{\Delta}(j\theta_k) = \begin{cases} \Delta & k = k_0 \\ \gamma I & k \in K \text{ and } k \neq k_0 \\ 0 & k \notin K \end{cases}.$$

Proof Set $\eta := \max\{\bar{\sigma}(\Delta), \gamma\}$. From Lemma 4.9 there exist scalar functions f_1 and f_2, both in $\mathcal{A}_{\mathbb{R}}$, so that

$$f_1(j\theta_k) = \begin{cases} 1 & k = k_0 \\ 0 & k \neq k_0 \end{cases} \quad ; \quad f_2(j\theta_k) = \begin{cases} 1 & k \in K \text{ and } k \neq k_0 \\ 0 & k \notin K \text{ or } k = k_0 \end{cases} .$$

Now, $f_2(j\theta_{k_0}) = 0$ and hence there exists an $\epsilon > 0$ so that on the interval $[-\epsilon + \theta_{k_0}, \epsilon + \theta_{k_0}]$ we have

$$f_2(j\omega) < \tfrac{1-\eta}{2}. \tag{4.4}$$

For $\alpha > 0$ we define $\hat{Q} \in \mathcal{A}_{\mathbb{R}}^{X}$ to be the function defined by

$$\begin{aligned} \hat{Q}(s) \quad := & \quad \left(\frac{2j\theta_{k_0} + 1/\alpha}{2j\theta_{k_0}}\right)\left(\frac{s + j\theta_{k_0}}{s + j\theta_{k_0} + 1/\alpha}\right)\left(\frac{\Delta}{1 + \alpha(s - j\theta_{k_0})}\right) \\ & + \left(\frac{2j\theta_{k_0} - 1/\alpha}{2j\theta_{k_0}}\right)\left(\frac{s - j\theta_{k_0}}{s - j\theta_{k_0} + 1/\alpha}\right)\left(\frac{(\Delta^*)^T}{1 + \alpha(s + j\theta_{k_0})}\right), \end{aligned}$$

where α is sufficiently large so that $\|\hat{Q}\|_\infty < \frac{1+\eta}{2}$ is fulfilled, and

$$\bar{\sigma}(\hat{Q}(j\omega)) < \frac{1-\eta}{2} \tag{4.5}$$

for $\omega \notin [-\epsilon + \theta_{k_0}, \epsilon + \theta_{k_0}] \cup [-\epsilon - \theta_{k_0}, \epsilon - \theta_{k_0}]$.

It is routine to verify that $\hat{\Delta} := f_1\hat{Q} + \gamma f_2 I$ meets the interpolation conditions required. To see that the norm constraint is also met, use first the following inequality which holds on the interval $[-\epsilon + \theta_{k_0}, \epsilon + \theta_{k_0}] \cup [-\epsilon - \theta_{k_0}, \epsilon - \theta_{k_0}]$:

$$\bar{\sigma}(\hat{Q}(j\omega))|f_1(j\omega)| + \gamma|f_2(j\omega)| < \frac{1+\eta}{2} + \gamma\frac{1-\eta}{2} < 1,$$

which follows by $\|\hat{Q}\|_\infty < \frac{1+\eta}{2}$, $\|f_1\|_\infty = \|f_2\|_\infty = 1$, and the inequality given in (4.4).

For ω not in the set $[-\epsilon + \theta_{k_0}, \epsilon + \theta_{k_0}] \cup [-\epsilon - \theta_{k_0}, \epsilon - \theta_{k_0}]$ we have that

$$\bar{\sigma}(\hat{Q}(j\omega))|f_1(j\omega)| + \gamma|f_2(j\omega)| < \frac{1-\eta}{2} + \gamma < 1,$$

from (4.5) and again that $\|f_1\|_\infty = \|f_2\|_\infty = 1$. Hence, by the last two inequalities we have that $\|\hat{\Delta}\|_\infty < 1$. ■

The following corollary to the last lemma is a key result used in the next section; it says that we can interpolate a finite number of matrices in $\mathcal{X}$ and an infinite number of zeros, along the imaginary axis, using a function in $\mathcal{A}_{\mathbb{R}}^{\mathcal{X}}$.

Corollary 4.11 *Suppose ω_0 is nonzero and in $(-\pi, \pi)$, and that θ_k is the infinite sequence $\{\frac{2\pi v_k - \omega_0}{h}\}_{k=0}^{\infty}$. Given a finite sequence of complex matrices $\{\underline{\Delta}_k\}_{k=0}^{n}$ in $\mathcal{X}$ with $\bar{\sigma}(\underline{\Delta}_k) < 1$, and a finite set of distinct non-negative integers $K = \{k_0, \ldots, k_n\}$, there exists $\hat{\Delta} \in \mathcal{UA}_{\mathbb{R}}^{\mathcal{X}}$ so that*

$$\hat{\Delta}(j\theta_k) = \begin{cases} \underline{\Delta}_k & k \in K \\ 0 & k \notin K \end{cases}.$$

Proof Let $\beta > 1$ be so that $\beta \max \bar{\sigma}(\underline{\Delta}_k) < 1$. By Lemma 4.10 there exist $n+1$ functions $\hat{\Delta}_l$, with $\|\hat{\Delta}_l\|_\infty < 1$, so that

$$\hat{\Delta}_l(j\theta_k) = \begin{cases} \beta\underline{\Delta}_k & k = k_l \\ \frac{1}{\sqrt[n]{\beta}} I & k \in K \text{ and } k \neq k_l \\ 0 & k \notin K \end{cases} \tag{4.6}$$

where $0 \le l \le n$. We get our desired function by setting

$$\hat{\Delta} := \Pi_{l=0}^{n} \hat{\Delta}_l.$$

It has norm less than one because $\|\hat{\Delta}_l\|_\infty < 1$ for $0 \le l \le n$, and by (4.6) interpolates the required matrices. ■

We have demonstrated that it is sufficient to verify a regularity condition on the unit circle (Lemma 4.5) in order to ensure robust stability, and have also determined the possible behavior of the LTI perturbations on this boundary (Corollary 4.11). We have now developed a foundation on which to construct the robustness criterion of the next section.

4.3 Robustness Test

Now we construct a necessary and sufficient robustness condition for the sampled-data stabilization problem we have posed. To accomplish this we change the problem domain by developing a new representation for the the transfer function $\check{M}(z)$ on the unit circle $\partial\mathbb{D}$, using the basis functions defined in (4.3). This new representation, the sampled-data frequency response, puts us in a position to more conveniently work with the results in Lemma 4.5 and Corollary 4.11, which leads to the construction of the stabilization condition.

To start, we have a lemma which states that the transfer function $\check{M}$ only takes values in the compact operators.

Lemma 4.12 *For each fixed $z_0 \in \bar{\mathbb{D}}$ the operator $\check{M}(z_0)$, defined in (3.5), is compact on $\mathcal{K}_2$.*

Proof Fix $z_0 \in \bar{\mathbb{D}}$. Now $\check{C} : \mathbb{C}^n \to \mathcal{K}_2$, making $\check{C}z_0(I - z_0A_d)^{-1}\check{B}$ finite rank and hence compact. The operator $\check{D}$ is an integral operator on $\mathcal{K}_2$, with a bounded kernel, which is well-known to be Hilbert-Schmidt and therefore compact. See for example [68, pp. 139]. ■

In fact, throughout the current section this is the only property of $\check{M}$ that is required for the analysis to remain valid. We therefore put the less restrictive assumption on $\check{M}$, namely that it is merely any function in $\mathcal{A}$ that maps $\partial\mathbb{D}$ to the compact operators $\mathfrak{C}(\mathcal{K}_2)$. Therefore, the results of this section apply to a wider class of h-periodic systems than just those arising from the linear fractional feedback connection of a continuous time LTI plant $\mathbf{G}$ and a sampled-data controller $\mathbf{HK}_d\mathbf{S}$ depicted in Figure 3.1.

At this point it is instructive to look at a picture. Fix $\omega_0 \in (-\pi, \pi]$ and consider the invertibility of the operator $I - \check{M}(e^{j\omega_0})\check{\Delta}(e^{j\omega_0})$ in terms of the well-posedness of the feedback shown in Figure 4.3. In the figure each $\Delta_k = \hat{\Delta}(j\frac{2\pi\nu_k - \omega_0}{h})$, and the operator $\check{M}(e^{j\omega_0})$ is considered as an infinite matrix, with entries M_{kl}, with respect to the basis ψ_k defined in (4.3). The block diagonal structure for $\check{\Delta}(e^{j\omega_0})$ follows from Corollary 4.8.

We can see clearly how the LTI nature of our perturbation Δ is captured by the fact that $\check{\Delta}$ is block diagonal. In the case of our initial problem having unstructured uncertainty, permitting $\check{\Delta}$ to be any function in $\mathcal{A}$ means

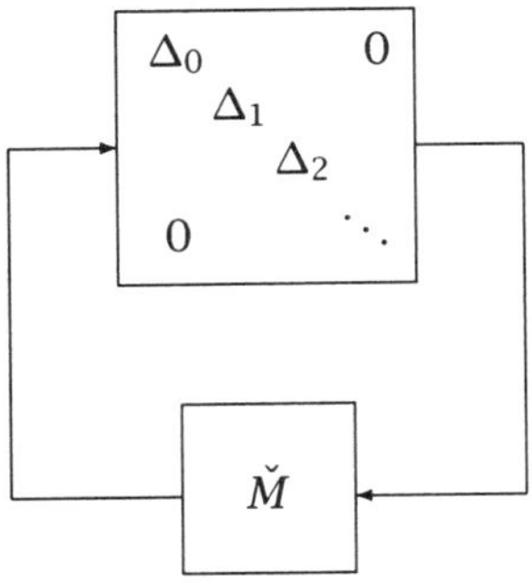

Figure 4.3: Stability Configuration

that $\check{\Delta}(e^{j\omega_0})$ could be any full-block matrix with respect to the basis ψ_k, rather than a block-diagonal one. Recall that enlarging the perturbation class to all of $\mathcal{A}$ corresponds to allowing $\boldsymbol{\Delta}$ to be periodically time-varying with period h instead.

Because of the structure of $I - \check{M}\check{\Delta}$ with respect to the particular basis ψ_k, it is more convenient to work with operators on the sequence space ℓ_2 rather than $\mathcal{K}_2$. We accomplish this transformation by first defining the set of operators $J_{\omega_0} : \mathcal{K}_2 \to \ell_2$ that are indexed by the parameter $\omega_0 \in (-\pi, \pi]$. For $\psi \in \mathcal{K}_2^m$ with Fourier expansion $\psi = \sum_{k=0}^{\infty} a_k \psi_k$, each $a_k \in \mathbb{C}^m$, the map is defined by

$$\sum_{k=0}^{\infty} a_k \psi_k \overset{J_{\omega_0}}{\longmapsto} (a_0, a_1, a_2, \ldots) \tag{4.7}$$

where ψ_k is the basis at $\omega_0 \in (-\pi, \pi]$ defined in (4.3).

We also regard the above set of maps, more generally, as a function mapping $\partial\mathbb{D}$ to $\mathfrak{L}(\mathcal{K}_2, \ell_2)$ via

$$J(e^{j\omega}) := J_\omega.$$

The inverse J^{-1} exists, since at each ω, we have that $J_\omega J_\omega^* = I$. It is therefore routine to see that J is an isomorphism, via multiplication, between the space of square integrable $\mathcal{K}_2$ valued functions on $\partial\mathbb{D}$ and the space of square integrable ℓ_2 valued functions on $\partial\mathbb{D}$.

We can now define the frequency response functions mapping the unit circle $\partial\mathbb{D}$ to the space of operators $\mathfrak{L}(\ell_2)$ by

$$\begin{aligned} M(e^{j\omega}) &:= J(e^{j\omega})\check{M}(e^{j\omega})J(e^{j\omega})^* \\ \Delta(e^{j\omega}) &:= J(e^{j\omega})\check{\Delta}(e^{j\omega})J(e^{j\omega})^*. \end{aligned} \tag{4.8}$$

We will examine the continuity properties of these functions more closely in Section 4.4.

From Corollary 4.8 it is clear that at every point on $\partial\mathbb{D}$ the perturbation $\Delta(e^{j\omega})$ is block diagonal. To make this precise define the subspace of $\mathfrak{L}(\ell_2)$

$$\boldsymbol{\Delta}_{LTI} := \{\operatorname{diag}(\Delta_0, \Delta_1, \Delta_2, \ldots) : \Delta_k \in \mathcal{X}\}. \tag{4.9}$$

Each element of this subspace is block diagonal, where the blocks take values in the original spatial uncertainty set $\mathcal{X}$ defined in (4.1). With this definition we have, at each point on the unit circle, that $\Delta(e^{j\omega})$ takes a value in $\boldsymbol{\Delta}_{LTI}$.

For the next lemma we require the projection operators, defined for each $n > 0$, $P_n : \ell_2 \to \ell_2$ defined by

$$(a_0, \ldots, a_n, a_{n+1}, \ldots) \overset{P_n}{\longmapsto} (a_0, \ldots, a_n, 0, 0, \ldots).$$

Note that P_n forms an increasing sequence of strongly converging projections. Also observe that from the above definitions, it is clear that any element $\Delta \in \boldsymbol{\Delta}_{LTI}$ commutes with P_n for any $n > 0$:

$$P_n\Delta = \Delta P_n = \operatorname{diag}(\Delta_0, \ldots, \Delta_n, 0 \ldots)$$

where $\Delta = \operatorname{diag}(\Delta_0, \Delta_1, \Delta_2, \ldots)$.

We now show that all the truncated elements of the set $\boldsymbol{\Delta}_{LTI}$ have a correspondence with the values of the frequency responses $\Delta(e^{j\omega})$. Recall that $\mathfrak{X}_{LTI}$ is the subspace of $\mathfrak{L}(\mathcal{L}_2)$ that corresponds to the structured transfer functions $\mathcal{A}_{\mathbb{R}}^{\mathcal{X}}$.

Lemma 4.13 *Suppose $\Delta' \in \mathcal{U}\boldsymbol{\Delta}_{LTI}$, $n > 0$, and that ω_0 is nonzero and is in the interval $(-\pi, \pi)$. Then there exists $\boldsymbol{\Delta} \in \mathcal{U}\mathfrak{X}_{LTI}$ so that $\Delta' P_n = \Delta(e^{j\omega_0})$.*

Proof We start by noting that $\Delta' P_n = \text{diag}(\Delta_0, \ldots, \Delta_n, 0, \ldots)$, with each element $\Delta_k \in \mathcal{X}$. By Corollary 4.11 we can construct a perturbation $\hat{\Delta} \in \mathcal{UA}_{\mathbb{R}}$ satisfying

$$\hat{\Delta}(j\tfrac{2\pi k - \omega_0}{h}) := \begin{cases} \Delta_k, & 0 \le k \le n \\ 0, & n < k \end{cases} .$$

Map $\hat{\Delta}$ to $\check{\Delta} \in \mathcal{U}\boldsymbol{\mathcal{A}}_{\mathcal{A}_{\mathbb{R}}}$ as in Lemma 4.1. Then from Corollary 4.8 and (4.8) we have that

$$\Delta(e^{j\omega_0}) = \text{diag}(\Delta_0, \ldots, \Delta_n, 0, \ldots) = \Delta' P_n. \quad \blacksquare$$

The next lemma gives a necessary condition for robust stability and is closely connected with the sufficient condition of Lemma 4.5.

Lemma 4.14 *Suppose $\Delta \in \mathcal{U}\boldsymbol{\Delta}_{LTI}$, that $\omega_0 \in (-\pi, \pi)$ and is nonzero. If rad($(M)(e^{j\omega_0})\Delta$) ≥ 1, then there exists $\boldsymbol{\Delta}' \in \mathcal{U}\boldsymbol{\mathfrak{X}}_{LTI}$ so that $I - (M\Delta')(e^{j\omega_0})$ is singular.*

Proof Because $\mathcal{U}\boldsymbol{\Delta}_{LTI}$ is an open set, there exists $1 > \beta > 0$ so that $(\frac{1}{1-\beta})\Delta \in \mathcal{U}\boldsymbol{\Delta}_{LTI}$.

By Lemmas 4.12 and Proposition 2.2 (i) we have that, for every $z \in \partial\mathbb{D}$, the operator $M(z)\Delta$ is compact. By Proposition 2.3 we have that $M(e^{j\omega_0})\Delta P_n \to M(e^{j\omega_0})\Delta$ as $n \to \infty$. Therefore by Proposition 2.3 (ii) there exists $n > 0$, sufficiently large, so that

$$|\text{rad}(M(e^{j\omega_0})\Delta) - \text{rad}(M(e^{j\omega_0})\Delta P_n)| < \beta. \tag{4.10}$$

Let $\lambda \in \text{spec}(M(e^{j\omega_0})\Delta P_n)$ so that $|\lambda| = \text{rad}(M(e^{j\omega_0})\Delta P_n)$. Since $\text{rad}(M(e^{j\omega_0})\Delta) \ge 1$ equation (4.10) ensures that

$$|\lambda| \ge 1 - \beta.$$

Hence, by our initial choice of β we have that $\lambda^{-1}\Delta \in \mathcal{U}\boldsymbol{\Delta}_{LTI}$. We also have that $1 \in \text{spec}(M(e^{j\omega})(\lambda^{-1}\Delta' P_n)\,)$. By Lemma 4.13 there exists $\Delta' \in$

$\mathcal{U}\mathfrak{X}_{LTI}$ so that the frequency response $\Delta'(e^{j\omega_0}) = \lambda^{-1}\Delta P_n$, and therefore $I - (M\Delta)(e^{j\omega_0})$ is singular. ■

The last lemma and Lemma 4.5 provide necessary and sufficient conditions for robust stabilization to perturbations in $\mathfrak{X}_{LTI}$. The theorem that follows is the main result of this chapter, and combines the two lemmas into a single necessary and sufficient condition for robust stability of the sampled-data system.

Theorem 4.15 *Given $r > 0$ and the previous definitions. Figure 4.1 has robust stability to $r\mathcal{U}\mathfrak{X}_{LTI}$ if and only if*

$$r \sup_{\omega \in (-\pi, \pi]} \mu_{\Delta_{LTI}}(M(e^{j\omega})) \leq 1. \tag{4.11}$$

The theorem states that the sampled-data system is robustly stabilized if and only if the structured singular value, defined in Section 2.6, of the frequency response $M(e^{j\omega})$ with respect to the uncertainty set $\boldsymbol{\Delta}_{LTI}$ satisfies the above inequality.

Proof Without loss of generality we set $r = 1$. From the definition of the structured singular value it is clear that $\sup_{\omega\in(-\pi,+\pi]} \mu_{\Delta_{LTI}}(M(e^{j\omega})) \leq 1$ if and only if for each $\Delta \in \mathcal{U}\boldsymbol{\Delta}_{LTI}$ the function $\text{rad}(M(e^{j\omega})\Delta) < 1$ on $\partial\mathbb{D}$. We work with this latter function.
(If:) By assumption we have that for each $\boldsymbol{\Delta} \in \mathcal{U}\mathfrak{X}_{LTI}$ the function $\text{rad}((M\Delta)(e^{j\omega})) < 1$ on $\partial\mathbb{D}$, then by Lemma 4.5 and Lemma 4.2 the system is robustly stabilized since $\mathcal{U}\mathfrak{X}_{LTI}$ is isomorphic to $\mathcal{U}\mathcal{A}_{\mathcal{A}_{\mathbb{R}}^{X}}$.
(Only if:) By contrapositive: suppose there exists a $\Delta \in \mathcal{U}\boldsymbol{\Delta}_{LTI}$ and an ω so that $1 > \text{rad}(M(e^{j\omega})\Delta)$. By the continuity result of Proposition 2.2 and Proposition 4.19, we may assume, without loss of generality, that ω is nonzero and not equal to π. Then from Lemma 4.14 and Lemma 4.2 we see immediately that the system cannot be robustly stabilized. ■

The above theorem is in terms of an infinite dimensional operator. With additional assumptions it is possible to convert this robust stability theorem to one in finite dimensions: the form of the sampled-data transfer function given in (3.5) is $\check{M}(e^{j\omega}) = \check{C}e^{j\omega}(I - e^{j\omega}A_d)^{-1}\check{B} + \check{D}$. From (4.8) we therefore define the RHS operators below

$$M(e^{j\omega}) =: \tilde{C}e^{j\omega}(I - e^{j\omega}A_d)^{-1}\tilde{B} + \tilde{D}, \tag{4.12}$$

where $\tilde{C} := J_\omega \check{C}$, $\tilde{B} := \check{B}J_\omega^*$, and $\tilde{D} := J_\omega \check{D} J_\omega^*$. The operator $\tilde{B}$ maps ℓ_2^m to $\mathbb{C}^{\tilde{n}}$, and can therefore be viewed as an infinite matrix

$$\tilde{B} =: [(\tilde{B})_0 \ (\tilde{B})_1 \ (\tilde{B})_2 \ldots],$$

where each $(\tilde{B})_k$ is defined to correspond to the matrix that acts on the kth component of a sequence in ℓ_2. Similarly, since $\tilde{C} : \mathbb{C}^{\tilde{n}} \to \ell_2^m$ and $\tilde{D} : \ell_2^m \to \ell_2^m$ they can also be regarded as infinite matrices, and we define their block entries by $(\tilde{C})_l$ and $(\tilde{D})_{lk}$ respectively. State space formulae for all these matrices can be found in Appendix A.

Now the additional assumption we impose to get a finite dimensional problem is that $\check{D} = 0$ in (3.5); or equivalently $\tilde{D} = 0$. In terms of the initial plant data this corresponds to the $\mathbf{G}_{11}$ block of the plant $\mathbf{G}$ being zero. The assumption therefore still allows multiplicative and additive uncertainty, but precludes more general types of linear fractional uncertainty. With the assumption we have that the frequency response $M(e^{j\omega})$ has the form

$$M(e^{j\omega}) = \tilde{C}e^{j\omega}(I - e^{j\omega}A_d)^{-1}\tilde{B}. \tag{4.13}$$

Having made this assumption we investigate an alternative form of the robustness test. Suppose that $\Delta \in \mathbf{\Delta}_{LTI}$ and $I - M(e^{j\omega})\Delta$ is singular. Then there exists a nonzero $\psi \in \ell_2$ so that $(I - M(e^{j\omega})\Delta)\psi = 0$. This is because $M(e^{j\omega})\Delta$ is compact and therefore its nonzero spectrum consists of eigenvalues; see for example [31]. So

$$(I - \tilde{C}e^{j\omega}(I - e^{j\omega}A_d)^{-1}\tilde{B}\Delta)\psi = 0.$$

This holds if and only if there exists a nonzero complex vector, say x, so that

$$(I - \tilde{B}\Delta\tilde{C}e^{j\omega}(I - e^{j\omega}A_d)^{-1})x = 0.$$

Define the sequence of matrices $\{\Delta_k\}_{k=0}^{\infty}$ in $\mathcal{X}$ so that $\Delta = \mathrm{diag}(\Delta_0, \Delta_1, \ldots)$, which is therefore compatible with the block structure $(\tilde{B})_k$ and $(\tilde{C})_k$ defined in (4.12). We therefore have that

$$(I - \sum_{k=0}^{\infty} (\tilde{B})_k \Delta_k (\tilde{C})_k e^{j\omega} (I - e^{j\omega} A_d)^{-1})x = 0. \tag{4.14}$$

Having made this observation, we have the following corollary to Theorem 4.15.

Corollary 4.16 *Suppose that $\check{D} = 0$ in (3.5). Given $r > 0$ and the previous definitions. Figure 4.1 has robust stability to $r\mathcal{U}\mathfrak{X}_{LTI}$ if and only if*

$$r \max_{\omega \in (-\pi, \pi]} \sup_{\Delta_k \in \mathcal{U}\mathcal{X}} \mathrm{rad}(\sum_{k=0}^{\infty} (\tilde{B})_k \Delta_k (\tilde{C})_k e^{j\omega} (I - e^{j\omega} \tilde{A})^{-1}) \leq 1.$$

This corollary provides a robustness condition in terms of a spectral radius maximization of a matrix of finite dimensions. It still however has an infinite number of uncertainty blocks.

The theorem developed in this section reduces checking robust stability of the sample-data system to a structured singular value test. This structured singular value is however that of an infinite dimensional operator; in the next chapter we develop a computational framework in which to address calculating this structured singular value.

In the section that follows we examine some of the elementary properties of the frequency response function.

4.4 Sampled-data Frequency Response

Our purpose here is to investigate some of the continuity properties of the frequency response function $M(e^{j\omega})$. This function is called the frequency response due to its connections with the continuous time sinusoidal functions $e^{j\omega_c t}$ as shown in Lemma 4.7. This connection was made in [25], and independently by Araki et. al [1]. Our interest in the frequency response is as a tool for analyzing robustness, and we therefore do not dwell on the above connection. Instead we examine the continuity of $M(e^{j\omega})$ which has implications for the behavior of $\mu_{\Delta_{LTI}}(M(e^{j\omega}))$.

The frequency response provides an infinite matrix representation of the transfer function $\check{M}$ on the unit circle. The transfer function is continuous on the unit circle, but $M(e^{j\omega})$ does not inherit this property; as we will see, except in special cases, it has a discontinuity introduced at one point.

Our first result shows that the operator-valued function J_ω or $J(e^{j\omega})$, defined in (4.7), is continuous on $(-\pi, \pi)$, and is left continuous at π:

Lemma 4.17

(i) If $\omega_0 \in (-\pi, \pi)$, *then* $\lim_{\omega\to\omega_0} \|J_{\omega_0} - J_\omega\|_{\mathcal{K}_2\to\ell_2} = 0$

(ii) $\lim_{\omega\uparrow\pi} \|J_\pi - J_\omega\|_{\mathcal{K}_2\to\ell_2} = 0.$

Proof We prove (i) and (ii) simultaneously. Let ω_0 be fixed in the interval $(-\pi, \pi]$. Now choose any function ψ in $\mathcal{K}_2$ with $\|\psi\|_{\mathcal{K}_2} = 1$, and $\omega \in (-\pi, \pi)$. Then by Proposition 4.6 this function has Fourier expansions

$$\psi(t) =: h^{-\frac{1}{2}} \sum_{k=0}^{\infty} a_k e^{j(\frac{2\pi v_k - \omega_0}{h})t} =: h^{-\frac{1}{2}} \sum_{k=0}^{\infty} b_k e^{j(\frac{2\pi v_k - \omega}{h})t}.$$

So

$$e^{j(\frac{\omega-\omega_0}{h})t}\psi(t) = h^{-\frac{1}{2}} \sum_{k=0}^{\infty} b_k e^{j(\frac{2\pi v_k - \omega_0}{h})t},$$

and hence we get that

$$\begin{aligned}\sum_{k=0}^{\infty} |a_k - b_k|_2^2 &= h\|\sum_{k=0}^{\infty} a_k e^{j(\frac{2\pi v_k-\omega_0}{h})t} - \sum_{k=0}^{\infty} b_k e^{j(\frac{2\pi v_k-\omega_0}{h})t}\|_{\mathcal{K}_2}^2 \\ &= \|(1 - e^{j(\frac{\omega-\omega_0}{h})t})\psi(t)\|_{\mathcal{K}_2}^2 \\ &\le \|(1 - e^{j(\frac{\omega-\omega_0}{h})t})\|_{\mathcal{K}_2}^2. \end{aligned} \tag{4.15}$$

Now by the definition of J_ω given in (4.7) we have that

$$J_{\omega_0}\psi - J_\omega\psi = (a_0 - b_0, a_1 - b_1, a_2 - b_2, \ldots).$$

Therefore using the inequality in (4.15) we get

$$\|J_{\omega_0}\psi - J_\omega \psi\|_{\ell_2} \le \|(1 - e^{j(\frac{\omega-\omega_0}{h})t})\|_{\mathcal{K}_2},$$

and because ψ is arbitrary we have that

$$\|J_{\omega_0} - J_\omega\|_{\mathcal{K}_2 \to \ell_2} \le \|(1 - e^{j(\frac{\omega-\omega_0}{h})t})\|_{\mathcal{K}_2}.$$

This inequality holds for any ω on the interval $(-\pi, \pi)$ and therefore the limits in (i) and (ii) follow immediately. ■

We now focus on the behavior of J_ω as ω tends to $-\pi$ from above. To accomplish this we define the operator $X : \ell_2 \to \ell_2$ via

$$(a_0, a_1, a_2, a_3, \dots) \overset{X}{\longmapsto} (a_2, a_0, a_4, a_1, a_6, a_3, a_8, \dots) \tag{4.16}$$

This operator simply reorders the elements in an ℓ_2 sequence. It is therefore unitary. We can now state and prove a corollary to the above lemma.

Corollary 4.18 *The following limit holds:*

$$\lim_{\omega \downarrow -\pi} \|XJ_\pi - J_\omega\|_{\mathcal{K}_2 \to \ell_2} = 0.$$

This corollary implies that $J(\cdot)$ is not continuous on $\partial\mathbb{D}$ since $J_\pi = J(-1)$. Intuitively, this discontinuity occurs because the basis defined in (4.3) has a discontinuity as ω tends to $-\pi$; this is shown in the proof.

Proof Choose any function ψ in $\mathcal{K}_2$ with $\|\psi\|_{\mathcal{K}_2} = 1$. Then by Proposition 4.6 this function has Fourier expansion

$$\psi(t) =: h^{-\frac{1}{2}} \sum_{k=0}^{\infty} a_k e^{j(\frac{2\pi \nu_k + \pi}{h})t}.$$

From this and the definition of X above

$$XJ_\pi\psi = (a_2, a_0, a_4, a_1, a_6, a_3, a_8, \dots) \tag{4.17}$$

holds.

Now by simply reordering the sequence we also have that

$$\psi(t) =: h^{-\frac{1}{2}} \sum_{k=0}^{\infty} a_{\eta_k} e^{j(\frac{2\pi v_k - \pi}{h})t},$$

where η_k is the sequence $\{2, 0, 4, 1, 6, 3, \ldots\}$. We also expand the function for some $\omega \in (-\pi, \pi]$ as

$$\psi(t) =: h^{-\frac{1}{2}} \sum_{k=0}^{\infty} b_k e^{j(\frac{2\pi v_k - \omega}{h})t}.$$

Then by the same steps as in Lemma 4.17 we can show that

$$\sum_{k=0}^{\infty} |a_{\eta_k} - b_k|_2 = \|(1 - e^{j(\frac{\omega+\pi}{h})t})\psi(t)\|_{\mathcal{K}_2}.$$

This holds for every ψ that has unit norm. Hence, from (4.17) and the definition of v_k we conclude that

$$\|XJ_\pi - J_\omega\|_{\mathcal{K}_2 \to \ell_2} \le \|(1 - e^{j(\frac{\omega+\pi}{h})t})\psi(t)\|_{\mathcal{K}_2}. \qquad \blacksquare$$

From the above two results, the fact that $\check{M}$ is in $\mathcal{A}$, and the definition of $M(e^{j\omega})$ we have immediately the following claim.

Proposition 4.19

(i) $M(e^{j\omega})$ *is continuous on* $(-\pi, \pi)$ *and left continuous at* π.

(ii) $\lim_{\omega \downarrow -\pi} \|M(e^{j\omega}) - XM(-1)X^*\|_{\ell_2 \to \ell_2} = 0.$

The final result of this section is connected to Proposition 4.19: it says that $\mu_{\Delta_{LTI}}(M(-1))$ and $\mu_{\Delta_{LTI}}(XM(-1)X^*)$ are equal; this is despite the fact that $M(-1)$ and $XM(-1)X^*$ are not equal. The proof amounts to making the observation that X is a permutation map of $\boldsymbol{\Delta}_{LTI}$ onto itself.

Lemma 4.20

$$\mu_{\Delta_{LTI}}(M(-1)) = \mu_{\Delta_{LTI}}(XM(-1)X^*)$$

Proof Choose any $\Delta = \text{diag}(\Delta_0, \Delta_1, \Delta_2, \ldots) \in \boldsymbol{\Delta}_{LTI}$. Then it is straightforward to verify that

$$X^*\Delta X = \text{diag}(\Delta_1, \Delta_3, \Delta_0, \Delta_5, \Delta_2, \Delta_7, \ldots) \in \boldsymbol{\Delta}_{LTI}. \tag{4.18}$$

Hence, $\mu_{\boldsymbol{\Delta}_{LTI}}(M(-1)) \geq \mu_{\boldsymbol{\Delta}_{LTI}}(XM(-1)X^*)$ since $I - XM(-1)X^*$ is singular if and only if $I - M(-1)X^*\Delta X$ is singular.

In the same way as above, we can show that if $\Delta' \in \boldsymbol{\Delta}_{LTI}$ then $X\Delta' X^* \in \boldsymbol{\Delta}_{LTI}$; therefore $\mu_{\boldsymbol{\Delta}_{LTI}}(M(-1)) \leq \mu_{\boldsymbol{\Delta}_{LTI}}(XM(-1)X^*)$ because $I - M(-1)\Delta'$ is singular if and only if $I - XM(-1)X^*(X\Delta' X^*)$ is. ■

The results contained here will be later utilized in showing various continuity properties of $\mu_{\boldsymbol{\Delta}_{LTI}}(M(e^{j\omega}))$ and its bounds.

4.5 Summary

We have constructed a mathematically precise condition for robust stabilization of sampled-data systems to structured LTI perturbations. This condition was obtained by first converting the initial time-domain robustness problem to an equivalent problem in the sampled-data frequency domain. By establishing a useful connection between the usual frequency domain and the sampled-data frequency domain, Corollary 4.8, we defined the sampled-data frequency response function. Our final robust stabilization condition, stated in Theorem 4.15, was then constructed in terms of a structured singular value of the frequency response function. This provides an explicit test for robustness, however computing this condition is still an issue. In the next chapter we concentrate on these computational issues.

For technical convenience we have worked with disc algebra perturbations, however the condition in Theorem 4.15 is also exact for $\mathcal{H}_\infty$ perturbations. The proof of this extension utilizes the properties stated in Propositions 2.13 and 2.14.

Earlier work on sampled-data robust stability to LTI perturbations can be found in Thompson, Stein and Athans [65], Thompson, Dailey and

Doyle [64], Sivashankar and Khargonekar [59], and Hara et al. [33]. All these papers consider unstructured uncertainty, and use small gain approaches to obtain sufficient robust stability conditions.

The sampled-data frequency response just developed is a general tool for analyzing sampled-data systems, and is studied in different sampled-data contexts in[1][2] and [69].

Chapter 5

A Computational Framework

The main robustness result of the previous chapter is a both necessary and sufficient condition for robust stabilization of a sampled-data system to structured LTI perturbations. The condition is given in terms of the structured singular value of the sampled-data system frequency response $\mu_{\Delta_{LTI}}(M(e^{j\omega}))$ evaluated on the unit circle. The capability to use this stability test therefore relies on the ability to effectively evaluate the structured singular value of the infinite dimensional operator $M(e^{j\omega})$. This chapter is aimed at developing a computational framework that addresses this task. The approach used is to obtain bounds: we construct a family of upper and lower bounds for $\mu_{\Delta_{LTI}}(M(e^{j\omega}))$ whose members converge to $\mu_{\Delta_{LTI}}(M(e^{j\omega}))$. The resulting computational procedure is one in which the accuracy of the bounds for $\mu_{\Delta_{LTI}}(M(e^{j\omega}))$ can be systematically improved at the cost of additional computational effort.

Again we frequently assume that the transfer function $\check{M}(z)$ is simply any element of $\mathcal{A}$ that takes values in the compact operators when evaluated on the closed unit disc. Hence the results are applicable to a wider class of systems than those arising from the arrangement in Figure 3.1.

5.1 Lower Bounds

Our goal is to provide a set of lower bounds for the structured singular value $\mu_{\Delta_{LTI}}(M(e^{j\omega}))$. We accomplish this by truncating the infinite dimensional operator $M(e^{j\omega})$ to obtain a set of finite dimensional matrices. The structured singular values of this set of matrices form a sequence of lower bounds for $\mu_{\Delta_{LTI}}(M(e^{j\omega}))$, which in the limit converges to the desired object $\mu_{\Delta_{LTI}}(M(e^{j\omega}))$.

To facilitate the development we define the finite rank operator Π_n^m which maps ℓ_2^m to $(\bigoplus_{k=0}^{n} \mathbb{C}^m)$. For $\psi = (a_0, a_1, \dots) \in \ell_2^m$ the map is defined

$$(a_0, a_1, \dots) \overset{\Pi_n^m}{\longmapsto} (a_0, \dots, a_n). \tag{5.1}$$

In the sequel we omit the spatial dimension m when referring to this operator. The adjoint of Π_n is signified by $(\Pi_n)^*$; note that it is the natural inclusion mapping of Euclidean space in ℓ_2. As defined, the maps are related to the projection operator of the last chapter by $(\Pi_n)^*\Pi_n = P_n$, which forms a sequence of increasing and strongly converging projections on ℓ_2.

We can now define the indexed set of truncated frequency responses by

$$\underline{M}_n(e^{j\omega}) := \Pi_n M(e^{j\omega})(\Pi_n)^*, \tag{5.2}$$

where $n \geq 0$. Hence $\underline{M}_n(e^{j\omega_0})$ is a submatrix of $M(e^{j\omega_0})$, at each $\omega_0 \in (-\pi, \pi]$, where $M(e^{j\omega_0})$ is viewed as an infinite matrix. Also, by Proposition 2.3, we see that $\underline{M}_n(e^{j\omega_0})$ tends to $M(e^{j\omega_0})$ in the limit as n tends to infinity. State space formula to construct the matrix $\underline{M}_n$ can be found in Appendix A.

Throughout this section, unless otherwise stated, we will assume that ω_0 is some fixed frequency in the interval $(-\pi, \pi]$, and will usually suppress the dependence of the various frequency responses on this fixed variable.

For each $n \geq 0$ it is easy to see that

$$\Pi_{n-1}(\Pi_n)^*\underline{M}_n\Pi_n(\Pi_{n-1})^* = \underline{M}_{n-1}.$$

That is, $\underline{M}_n$ for fixed n, contains all $\underline{M}_q$ as submatrices, where $0 \le q \le n$.

Associated with the matrices $\underline{M}_n$ we define the uncertainty sets of matrices

$$\underline{\mathbf{\Delta}}_n := \{\mathrm{diag}(\underline{\Delta}_0, \ldots, \underline{\Delta}_n) : \underline{\Delta}_k \in \mathcal{X}\},$$

for each $n \ge 0$. Clearly they are the result of truncations of the elements in the set $\mathbf{\Delta}_{LTI}$; hence for each $n \ge 0$ the structured singular value $\mu_{\underline{\mathbf{\Delta}}_n}(\underline{M}_n)$ is well-defined. We have the following theorem.

Theorem 5.1 *The following statements hold:*

(i) $\mu_{\underline{\mathbf{\Delta}}_{n-1}}(\underline{M}_{n-1},) \le \mu_{\underline{\mathbf{\Delta}}_n}(\underline{M}_n) \le \mu_{\mathbf{\Delta}_{LTI}}(M)$, *for each* $n \ge 1$.

(ii) $\lim_{n\to\infty} \mu_{\underline{\mathbf{\Delta}}_n}(\underline{M}_n) = \mu_{\mathbf{\Delta}_{LTI}}(M)$.

The result states that the sequence of structured singular values $\mu_{\underline{\mathbf{\Delta}}_n}(\underline{M}_n)$ is monotonically increasing and forms a set of lower bounds for the sampled-data structured singular value $\mu_{\mathbf{\Delta}_{LTI}}(M)$. Furthermore, this sequence converges to $\mu_{\mathbf{\Delta}_{LTI}}(M)$.

Proof We start by proving $\mu_{\underline{\mathbf{\Delta}}_n}(\underline{M}_n) \le \mu_{\mathbf{\Delta}_{LTI}}(M)$ of (i). We will show that if $\underline{\Delta} \in \underline{\mathbf{\Delta}}_n$ and $I - \underline{M}_n\underline{\Delta}$ is singular then $I - M\Pi_n^*\underline{\Delta}\Pi_n$ is also singular; this is sufficient because $\|\Pi_n^*\underline{\Delta}\Pi_n\| = \bar{\sigma}(\underline{\Delta})$ and $\Pi_n^*\underline{\Delta}\Pi_n \in \mathbf{\Delta}_{LTI}$.

To start let $\underline{\Delta}$ be such a perturbation and let x be a nonzero vector in the kernel of $I - \underline{M}_n\underline{\Delta}$. Now $I - M\Pi_n^*\underline{\Delta}\Pi_n$ has the form

$$\begin{bmatrix} I & 0 \\ 0 & I \end{bmatrix} - \begin{bmatrix} \underline{M}_n & - \\ Q & - \end{bmatrix} \begin{bmatrix} \underline{\Delta} & 0 \\ 0 & 0 \end{bmatrix}, \tag{5.3}$$

where Q is some linear operator and $-$ denotes irrelevant operators. From (5.3) it is easy to see that $\begin{bmatrix} x \\ Q\underline{\Delta}x \end{bmatrix}$ is in the kernel of $I - M\Pi_n^*\underline{\Delta}\Pi_n$; hence the operator is singular.

To show the second inequality, a similar argument can be used by recalling that $\underline{M}_{n-1}$ is a submatrix of $\underline{M}_n$.

To start the proof of (ii) note that the limit exists and is bounded by $\mu_{\mathbf{\Delta}_{LTI}}(M)$ from (i). We will show that $\lim_{n\to\infty} \mu_{\underline{\mathbf{\Delta}}_n}(\underline{M}_n) \ge \mu_{\mathbf{\Delta}_{LTI}}(M)$.

Choose any $\epsilon > 0$ and let $\Delta \in \mathcal{U}\boldsymbol{\Delta}_{LTI}$ be so that

$$\epsilon/2 + \text{rad}(M\Delta) > \mu_{\boldsymbol{\Delta}_{LTI}}(M). \tag{5.4}$$

By Propositions 2.2 and 2.3 there exists n sufficiently large so that $\epsilon/2 + \text{rad}(P_n M\Delta P_n) > \text{rad}(M\Delta)$, and therefore by (5.4) we have that

$$\epsilon + \text{rad}(P_n M\Delta P_n) > \mu_{\boldsymbol{\Delta}_{LTI}}(M). \tag{5.5}$$

Now, from the diagonal structure of Δ it is straightforward to verify that $\Delta P_n = P_n\Delta$; since P_n is a projection we have that $\Delta P_n = \Delta P_n^2 = P_n\Delta P_n$. So

$$\begin{aligned} P_n M\Delta P_n &= P_n M P_n \Delta P_n \\ &= \Pi_n^*(\Pi_n M \Pi_n^*)(\Pi_n \Delta \Pi_n^*)\Pi_n. \end{aligned}$$

From the RHS we can see that the LHS above is isomorphic to the matrix $(\Pi_n M \Pi_n^*)(\Pi_n \Delta \Pi_n^*)$. Therefore it follows that $\text{rad}(P_n M\Delta P_n) = \text{rad}(\underline{M}_n \Pi_n \Delta \Pi_n^*)$, where we have also used the definition in (5.2). Now, $\bar{\sigma}(\Pi_n \Delta \Pi_n^*) < \|\Delta\| < 1$ and $(\Pi_n \Delta \Pi_n^*) \in \underline{\boldsymbol{\Delta}}_n$, therefore by definition $\mu_{\underline{\boldsymbol{\Delta}}_n}(\underline{M}_n) \geq \text{rad}(P_n M\Delta P_n)$. The claim then follows by (5.5). ■

A direct result of the theorem is the following result which relates back to the sampled-data system stability; it follows by invoking Theorem 4.15.

Corollary 5.2 *Given $r > 0$ and the previous definitions. Figure 4.1 has robust stability to $r\mathcal{U}\boldsymbol{\mathfrak{X}}_{LTI}$ if and only if for each $n \geq 0$ the inequality*

$$r \sup_{\omega \in (-\pi, +\pi]} \mu_{\underline{\boldsymbol{\Delta}}_n}(\underline{M}_n(e^{j\omega})) \leq 1 \qquad \textit{is satisfied.} \tag{5.6}$$

This corollary provides a necessary and sufficient condition for robust stabilization in terms of a sequence of necessary conditions. It is therefore not usable as a sufficient test for robust stability since only a finite number of computations can be performed. However, under restrictive conditions this test can be directly extended to provide sufficient conditions as well.

We accomplish this by making the same assumption as in Corollary 4.16.

Proposition 5.3 *Suppose that $\check{D} = 0$ in (3.5). Then for each $n \geq 0$*

$$\mu_{\underline{\Delta}_n}(\underline{M}_n(e^{j\omega})) = \sup_{\Delta_k \in \mathcal{U}X} \text{rad}\left(\sum_{k=0}^{n} (\tilde{B})_k \Delta_k (\tilde{C})_k \, e^{j\omega}(I - e^{j\omega} A_d)^{-1} \right).$$

Comparing the above result with Corollary 4.16 we see that the effect of truncating $M(e^{j\omega})$ by projection has the effect of truncating the tail of the infinite series. The assumption in the claim means that the frequency response has the form given in (4.13).

Proof Suppose that $\underline{\Delta} = \text{diag}(\Delta_0, \ldots, \Delta_n) \in \underline{\mathbf{\Delta}}_n$. Substituting the form of $M(e^{j\omega})$ in (4.12), and the definition $\underline{M}_n(e^{j\omega})$ in (5.2) we have that

$$\begin{aligned} I - \underline{M}_n(e^{j\omega})\underline{\Delta} &= I - \Pi_n M (\Pi_n)^* \underline{\Delta} \\ &= I - \Pi_n (\tilde{C} e^{j\omega}(I - e^{j\omega}\tilde{A})^{-1}\tilde{B})(\Pi_n)^* \underline{\Delta}. \end{aligned}$$

It is straightforward to see that this operator is singular if and only if the matrix

$$I - \tilde{B}(\Pi_n)^* \underline{\Delta} \Pi_n \tilde{C} \, e^{j\omega}(I - e^{j\omega}\tilde{A})^{-1} = I - \sum_{k=0}^{n} (\tilde{B})_k \Delta_k (\tilde{C})_k e^{j\omega}(I - e^{j\omega}\tilde{A})^{-1}$$

is singular, where the RHS follows simply from the definitions of (4.12). Since $\bar{\sigma}(\underline{\Delta}) = \max_{0 \leq k \leq n} \bar{\sigma}(\Delta_k)$ the result follows. ∎

Now the formulation of the lower bound $\mu_{\underline{\Delta}_n}(\underline{M}_n(e^{j\omega}))$ as the optimization over a series in Proposition 5.3 does not seem to represent a significant improvement in terms of computation. This can be seen from the work of Oishi [45], who studied a special case of this test in which the matrices $(\tilde{B})_k$ are scaled identities and the spatial uncertainty set X is unstructured. In order to carry out this optimization a nonconvex optimization was required, with no guarantee of convergence.

The advantage of the formulation in the proposition is that, as n increases, the rank of the resulting sum remains bounded by the constant dimension of the matrix A_d; this means it is possible to obtain (conservative) bounds on how much the spectral radius of a finite sum can vary as additional terms are added. We now develop some elementary results to this end.

To derive bounds on the variation of the spectral radius, we use a general result found in Elsner [26] on the variation of the spectrum of matrices. Given two square complex matrices E and F of equal size $q \times q$, [26] provides a result that specializes to the following inequality

$$\mathrm{rad}(E) - \mathrm{rad}(F) \leq (2m_{EF})^{(1-1/q)}(q\bar{\sigma}(E-F))^{1/q}, \tag{5.7}$$

where $m_{EF} := \max\{\bar{\sigma}(E), \bar{\sigma}(F)\}$. The paper gives a number of other interesting results that provide specific bounds on variation of the spectrum of two matrices; the one in (5.7) is the best suited for our purposes in the next result.

Proposition 5.4 *Suppose that $\check{D} = 0$ in (3.5). Then for each $n \geq 0$*

$$\mu_{\underline{\Delta}_n}(\underline{M}_n(e^{j\omega})) \leq \mu_{\Delta_{LTI}}(M(e^{j\omega})) \leq \mu_{\underline{\Delta}_n}(\underline{M}_n(e^{j\omega})) + \epsilon_n,$$

where $\epsilon_n = (2r_0)^{1-1/\tilde{n}}(\tilde{n}r_{n+1})^{1/\tilde{n}}$, the dimension of A_d is $\tilde{n} \times \tilde{n}$, and

$$r_n := \sum_{k=n+1}^{\infty} \bar{\sigma}((\tilde{B})_k)\bar{\sigma}((\tilde{C})_k\, e^{j\omega}(I - e^{j\omega}A_d)^{-1}).$$

The proposition states that it is possible to bound the structured singular value $\mu_{\Delta_{LTI}}(M(e^{j\omega}))$ between its lower bound $\mu_{\underline{\Delta}_n}(\underline{M}_n(e^{j\omega}))$ and its lower bound plus a quantity ϵ_n. As n increases, ϵ_n tends to zero and therefore this result provides a way of obtaining $\mu_{\Delta_{LTI}}(M(e^{j\omega}))$ to any desired degree of accuracy in terms of its lower bound. Appendix B provides a proof of this result.

The quantity ϵ_n given in Proposition 5.4 will typically be a conservative estimate for the size of the band in which $\mu_{\Delta_{LTI}}(M(e^{j\omega}))$ lies. This is because the spectral variation results in [26] are usually conservative; however they are apparently the best general bounds currently available.

The main objective here was to provide a converging set of lower bounds for the structured singular value $\mu_{\Delta_{LTI}}(M)$; this was accomplished by forming a truncated structured singular value problem on a finite matrix as given in Theorem 5.1.

In addition, we also provided some partial results on converging upper bounds, which are applicable in some additive and multiplicative uncertainty arrangements. It was possible to obtain these bounds by making assumptions that guaranteed the operator $M(e^{j\omega})$ be finite rank. Next we develop a general method for obtaining upper bounds for $\mu_{\Delta_{LTI}}(M)$ that tackles the infinite dimensional rank of $M(e^{j\omega})$ directly.

5.2 Upper Bounds

Our emphasis now is on deriving a set of computable sufficient conditions that guarantee robust stabilization of the sampled-data system. We achieve this by developing a set of upper bounds for $\mu_{\Delta_{LTI}}(M)$.

Our approach is to initially expand the structured uncertainty set, relaxing some of the constraints that are imposed on it by membership in Δ_{LTI}. The benefit of this step is that the problem of computing the structured singular value of M with respect to the enlarged uncertainty class can be reduced to a finite dimensional computation. Furthermore, we can choose the uncertainty class so that the resulting structured singular value is as close as desired to $\mu_{\Delta_{LTI}}(M)$.

The section is divided into three parts: after defining the upper bound, we show that it converges to $\mu_{\Delta_{LTI}}(M)$ monotonically; the only assumption that we make here about $M(e^{j\omega})$ is that it is a compact operator. In 5.2.2 we demonstrate how the infinite dimensional modified problem can be reduced to a computation in finite dimensions. In order to use this finite dimensional test a number of matrices are required, and the final subsection provides a technique to obtain them explicitly.

We first decompose the operator $M(e^{j\omega_0})$. To carry this out, fix a frequency ω_0 in the interval $(-\pi, \pi]$ and define the map $\Gamma_n^m : \ell_2^m \to \ell_2^m$ to be that which takes an element $\psi = (a_0, a_1, \ldots) \in \ell_2^m$ to

$$\Gamma_n^m \psi := (a_{n+1}, a_{n+2}, \ldots).$$

Hence the map $\Upsilon_n^m : \ell_2^m \to (\bigoplus_{k=0}^{n} \mathbb{C}^m) \oplus \ell_2^m$ defined by

$$(\Upsilon_n^m)^* := [(\Pi_n^m)^* \quad (\Gamma_n^m)^*]$$

is an isomorphism between the spaces, where Π_n^m was defined in (5.1) of the last section.

With these maps we can now decompose the operator $M(e^{j\omega_0})$ by setting

$$M_n(e^{j\omega_0}) := \Upsilon_n^m M(e^{j\omega_0})(\Upsilon_n^m)^* =: \begin{bmatrix} M_n^{11} & M_n^{12} \\ M_n^{21} & M_n^{22} \end{bmatrix}.$$

This decomposition is simply a partitioning of M when viewed as an infinite matrix; note that $M_n^{11} = \underline{M}_n$, where the RHS is the matrix associated with the lower bound of the previous section. As usual we drop the dependence of M, M_n and $\underline{M}_n$ on the fixed value $e^{j\omega_0}$; however it is always implied.

Define the following enlarged class of perturbations which are compatible with M_n,

$$\boldsymbol{\Delta}_n := \{\Delta = \text{diag}(\Delta^1, \Delta^2) : \Delta^1 \in \underline{\boldsymbol{\Delta}}_n, \Delta^2 \in \mathfrak{L}(\ell_2^m)\}.$$

Therefore the structured singular value $\mu_{\boldsymbol{\Delta}_n}(M_n)$ is well-defined for all $n \geq 0$. We further define $\boldsymbol{\Delta}_{-1} := \mathfrak{L}(\ell_2^m)$ and therefore $\mu_{\boldsymbol{\Delta}_{-1}}(M_{-1}(e^{j\omega})) = \|M(e^{j\omega})\|$. We also suppress these spatial dimension m in Π_n and Υ_n.

Associated with this uncertainty set is Figure 5.1 which contrasts with the setup of Figure 4.3: the enlarged uncertainty set Δ_n allows an infinite full-block perturbation, whereas the corresponding perturbation in Figure 4.3 must be block-diagonal. Thus the new perturbation class amounts to removing the infinite number of constraints put on an element in $\boldsymbol{\Delta}_{LTI}$; and it only imposes a diagonal constraint on the first $n+1$ blocks.

The next lemma states that $\mu_{\boldsymbol{\Delta}_n}(M_n)$ is an upper bound for $\mu_{\boldsymbol{\Delta}_{LTI}}(M)$.

Lemma 5.5 *For any non-negative integers n the following inequalities hold*

$$\mu_{\boldsymbol{\Delta}_{LTI}}(M) \leq \mu_{\boldsymbol{\Delta}_n}(M_n) \quad \textit{and} \quad \mu_{\boldsymbol{\Delta}_{n+1}}(M_{n+1}) \leq \mu_{\boldsymbol{\Delta}_n}(M_n).$$

Proof To obtain the first inequality: this follows directly from the fact that for any $\Delta \in \boldsymbol{\Delta}_{LTI}$ we have that

$$\Upsilon_n \Delta (\Upsilon_n)^* =: \Delta_n \in \boldsymbol{\Delta}_n,$$

and $I - M\Delta$ singular implies that $I - M_n \Delta_n$ is singular.

Monotonicity of $\mu_{\boldsymbol{\Delta}_n}(M_n)$ follows from the fact that for each $\Delta_{n+1} \in \boldsymbol{\Delta}_{n+1}$ we have $\Upsilon_n \Upsilon_{n+1}^* \Delta_{n+1} \Upsilon_{n+1} \Upsilon_n^* \in \boldsymbol{\Delta}_n$. ■

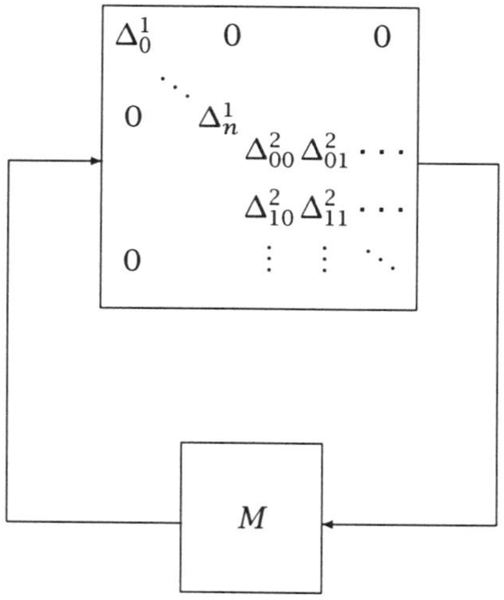

Figure 5.1: Modified Configuration

Next subsection we will show that the infinite sequence $\mu_{\boldsymbol{\Delta}_n}(M_n)$ converges to $\mu_{\boldsymbol{\Delta}_{LTI}}(M)$; from the last lemma this convergence is monotonic.

5.2.1 Convergence

The major aim is to prove that the set of upper bounds $\mu_{\boldsymbol{\Delta}_n}(M_n)$ has elements as close as desired to the structured singular value $\mu_{\boldsymbol{\Delta}_{LTI}}(M)$. We also prove some results regarding the continuity properties of $\mu_{\boldsymbol{\Delta}_{LTI}}(\cdot)$ and its related bounds. We require some technical lemmas to get started.

Lemma 5.6 *Suppose that Δ_k is a bounded sequence in $\mathfrak{L}(\ell_2)$. Then there exists a subsequence Δ_{k_n} and $\Delta \in \mathfrak{L}(\ell_2)$ so that*

$$\|P_n \Delta_{k_n} P_n - P_n \Delta P_n\| \overset{n\to\infty}{\longrightarrow} 0, \quad \textit{where } P_n := \Pi_n^* \Pi_n.$$

Furthermore, the limit $\lim_{n\to\infty} \|\Delta_{k_n}\| \geq \|\Delta\|$.

The lemma gives a compactness type result for sequences in $\mathfrak{L}(\ell_2)$, where finite dimensional projections of the opertors are taken. A consequence of the lemma is also that $P_n \Delta_{k_n}$ tends strongly to Δ as $n \to \infty$.

Proof Immediately we have by Alaoglu's theorem, see for example [11], that there is a subsequence of Δ_k that converges weak* to some element Δ in $\mathfrak{L}(\ell_2)$. Without loss of generality we assume that Δ_k converges weak* to Δ.

Fix n. By the definition of weak* convergence, we have that for every $u, v \in \ell_2$ that

$$\langle \Delta_k P_n u, P_n v \rangle \overset{k\to\infty}{\longrightarrow} \langle \Delta P_n u, P_n v \rangle, \tag{5.8}$$

since $P_n u$ and $P_n v$ are elements in ℓ_2. We next exploit the relationship

$$\begin{aligned} \langle \Delta_k P_n u, P_n v \rangle &= \langle \Delta_k \Pi_n^* \Pi_n u, \Pi_n^* \Pi_n v \rangle \\ &= \langle (\Pi_n \Delta_k \Pi_n^*) \Pi_n u, \Pi_n v \rangle. \end{aligned}$$

Similarly we get that $\langle \Delta P_n u, P_n v \rangle = \langle (\Pi_n \Delta \Pi_n^*) \Pi_n u, \Pi_n v \rangle$, and therefore by (5.8) and linearity that

$$\langle (\Pi_n \Delta_k \Pi_n^* - \Pi_n \Delta \Pi_n^*) \Pi_n u, \Pi_n v \rangle \overset{k\to\infty}{\longrightarrow} 0 \quad \text{for every } u, v \in \ell_2. \tag{5.9}$$

Now, for each k the map $(\Pi_n \Delta_k \Pi_n^* - \Pi_n \Delta \Pi_n^*)$ is just a bounded linear operator on the Euclidean space $(\bigoplus_{k=0}^{n} \mathbb{C}^p)$. Also, $\Pi_n u$ and $\Pi_n v$ can take any desired values in $(\bigoplus_{k=0}^{n} \mathbb{C}^p)$ by varying u and v. Hence, from (5.9) we conclude that

$$\|\Pi_n \Delta_k \Pi_n^* - \Pi_n \Delta \Pi_n^*\| \overset{k\to\infty}{\longrightarrow} 0.$$

Since Π_n^* is an injective isometry we know that $\|\Pi_n \Delta_k \Pi_n^* - \Pi_n \Delta \Pi_n^*\| = \|P_n \Delta_k P_n - P_n \Delta P_n\|$, and therefore $\|P_n \Delta_k P_n^* - P_n \Delta P_n^*\| \overset{k\to\infty}{\longrightarrow} 0$. For each n we can choose k_n large enough so that $\|P_n \Delta_{k_n} P_n - P_n \Delta P_n\| < (\frac{1}{2})^n$.

We now show the norm inequality: first, without loss of generality we may assume that the limit $\lim_{n\to\infty} \|\Delta_{k_n}\|$ exists since $\|\Delta_{k_n}\|$ is a bounded sequence. Choose $\epsilon > 0$ and let x, $y \in \ell_2$ be so that $\|x\|_2 = \|y\|_2 = 1$ and

$$\|\Delta\| \le \langle y, \Delta x\rangle_2 + \epsilon.$$

Then by weak* convergence and the Cauchy-Schwartz inequality we have for n sufficiently large that the following equalities hold

$$\|\Delta\| \;\le\; |\langle y, \Delta_{k_n} x\rangle_2| + \epsilon \;\le\; \|\Delta_{k_n} x\|_2 + \epsilon \;\le\; \|\Delta_{k_n}\| + \epsilon. \quad \blacksquare$$

We can now prove the next result using the above lemma, where we remove the projection operators to get norm convergence.

Lemma 5.7 *Suppose that E is a compact operator in $\mathfrak{L}(\ell_2)$, and that Δ_k is a sequence on the unit sphere of $\mathfrak{L}(\ell_2)$ satisfying $\Upsilon_k \Delta_k \Upsilon_k^* \in \boldsymbol{\Delta}_k$ for each k. Then there exist $\Delta \in \boldsymbol{\Delta}_{LTI}$, with $\|\Delta\| \le 1$ and a subsequence Δ_{k_n} satisfying*

$$\|E\Delta_{k_n} - E\Delta\| \overset{n\to\infty}{\longrightarrow} 0.$$

The lemma associates a single perturbation with every sequence Δ_k in $\boldsymbol{\Delta}_k$.

Proof We start by observing that the block diagonal structure imposed by $\Upsilon_k \Delta_k \Upsilon_k^* \in \boldsymbol{\Delta}_k$ implies that $P_n \Delta_{k_n} P_n = P_n \Delta_{k_n}$. Hence from Lemma 5.6 we have

$$\|P_n \Delta_{k_n} - P_n \Delta P_n\| \overset{n\to\infty}{\longrightarrow} 0. \tag{5.10}$$

This implies that $\|\Pi_n \Delta_{k_n} \Pi_n^* - \Pi_n \Delta \Pi_n^*\| \overset{n\to\infty}{\longrightarrow} 0$; therefore given any fixed n we have that $\|\Pi_n \Delta_{k_l} \Pi_n^* - \Pi_n \Delta \Pi_n^*\| \overset{l\to\infty}{\longrightarrow} 0$. For every $l \ge n$ we have that $\Pi_n \Delta_{k_l} \Pi_n^* \in \underline{\boldsymbol{\Delta}}_n$, because $\Upsilon_k \Delta_k \Upsilon_k^* \in \boldsymbol{\Delta}_k$ for each k. So,

$$\Pi_n \Delta \Pi_n^* \in \underline{\boldsymbol{\Delta}}_n,$$

for every n. This is true if and only if $\Delta \in \mathbf{\Delta}_{LTI}$.

Next we note that since E is compact so is $E\Delta$. Hence, by Proposition 2.2 (ii) it follows that $\|E - P_n E P_n\| \overset{n\to\infty}{\longrightarrow} 0$ and $\|E\Delta - P_n E P_n \Delta P_n\| \overset{n\to\infty}{\longrightarrow} 0$. Using the last two limits, (5.10), the triangle inequality, and the submultiplicative inequality, it is straightforward to demonstrate that $\|E\Delta_{k_n} - E\Delta\| \overset{n\to\infty}{\longrightarrow} 0$. ■

The following corollary follows directly from the continuity of the norm.

Corollary 5.8 *Given the supposition of Lemma 5.7. If E_k is a sequence in $\mathfrak{L}(\ell_2)$ tending to E, then there exist $\Delta \in \mathbf{\Delta}_{LTI}$, with $\|\Delta\| \leq 1$, and a subsequence Δ_{k_n} satisfying*

$$\|E_{k_n}\Delta_{k_n} - E\Delta\| \overset{n\to\infty}{\longrightarrow} 0.$$

Proof Choose Δ_{k_n} and Δ as in Lemma 5.7. Now use the inequality

$$\|E_{k_n}\Delta_{k_n} - E\Delta\| \leq \|E_{k_n} - E\| \cdot \|\Delta_{k_n}\| + \|E\Delta_{k_n} - E\Delta\|,$$

which follows from the triangle and submultiplicative inequalities. As $n \to \infty$ the LHS$\to 0$ by Lemma 5.7 and the assumption that E_{k_n} converges to E. ■

We are now in a position to prove the main theorem:

Theorem 5.9 *The following holds*

$$\lim_{n\to\infty} \mu_{\Delta_n}(M_n) = \mu_{\Delta_{LTI}}(M).$$

The theorem states the sequence of upper bounds $\mu_{\Delta_n}(M_n)$ converges to the sampled-data structured singular value. Note that by Lemma 5.5 this convergence is monotonic.

Proof It is sufficient to demonstrate that a subsequence $\mu_{\Delta_{n_k}}(M_{n_k})$ converges to $\mu_{\Delta_{LTI}}(M)$ since by Lemma 5.5 the sequence $\mu_{\Delta_n}(M_n)$ is monotonically decreasing.

By the definition of $\mu_{\Delta_n}(M_n)$ and the isomorphism Υ_n, we have that there exist Δ_n on the unit sphere of $\mathfrak{L}(\ell_2)$ each satisfying

$$\Upsilon_n \Delta_n \Upsilon_n^* \in \mathbf{\Delta}_n \quad \text{and} \tag{5.11}$$

$$\text{rad}(M\Delta_n) \le \mu_{\Delta_n}(M_n) \le (\tfrac{1}{2})^n + \text{rad}(M\Delta_n). \tag{5.12}$$

Since M is compact, by Lemma 5.7 there exist $\Delta \in \mathbf{\Delta}_{LTI}$, with $\|\Delta\| \le 1$ and a subsequence Δ_{k_n} satisfying

$$\|M\Delta_{k_n} - M\Delta\| \overset{n\to\infty}{\longrightarrow} 0. \tag{5.13}$$

Now by (5.12) and the definition of the structured singular value we get that for every n,

$$\text{rad}(M\Delta) \le \mu_{\Delta_{LTI}}(M) \le \mu_{\Delta_n}(M_{k_n}) \le (\tfrac{1}{2})^n + \text{rad}(M\Delta_{k_n}).$$

But by (5.13) and the continuity of the spectral radius of compact operators, Proposition 2.2, we have that $\text{rad}(M\Delta_{k_n})$ tends to $\text{rad}(M\Delta)$ as $n \to \infty$. That is, LHS=$\lim_{n\to\infty}$RHS above. ■

We have now shown that the structured singular value $\mu_{\Delta_{LTI}}(M(e^{j\omega_0}))$ can be bounded to any desired degree of accuracy between $\mu_{\Delta_n}(M_n(e^{j\omega_0}))$ and $\mu_{\underline{\Delta}_n}(\underline{M}_n(e^{j\omega_0}))$, *at each* $\omega_0 \in (-\pi, \pi]$. However, to use Theorem 4.15 we must be able to evaluate $\mu_{\Delta_{LTI}}(M(z))$ on the unit circle $\partial\mathbb{D}$; therefore our bounds must have the property that they uniformly approximate this structured singular value on the unit circle if they are to be of practical value. Our present goal is to establish that this is the case.

First we must prove some continuity results about the structured singular value.

Proposition 5.10 *Suppose that E is a compact operator on ℓ_2. Then the structured singular value function $\mu_{\Delta_{LTI}}(\cdot)$, mapping $\mathfrak{L}(\ell_2)$ to $\mathbb{R}$, is continuous at E.*

Proof Our proof shows first upper, and then lower, semicontinuity of the function $\mu_{\Delta_{LTI}}(\cdot)$.

To prove upper semicontinuity, we assume that $\mu_{\mathbf{\Delta}_{LTI}}(\cdot)$ does not have this property and derive a contradiction: suppose $\mu_{\mathbf{\Delta}_{LTI}}(\cdot)$ is not upper semicontinuous. Then there exists a sequence E_k tending to E, and a real number $\beta > 0$ so that

$$\mu_{\mathbf{\Delta}_{LTI}}(E) < \beta \le \mu_{\mathbf{\Delta}_{LTI}}(E_k), \quad \text{for all } k \ge 0. \tag{5.14}$$

Let Δ_k be a sequence in $\mathbf{\Delta}_{LTI}$ so that $\|\Delta_k\| = 1$ and

$$\operatorname{rad}(E_k\Delta_k) \le \mu_{\mathbf{\Delta}_{LTI}}(E_k) \le \operatorname{rad}(E_k\Delta_k) + (\tfrac{1}{2})^n. \tag{5.15}$$

By Corollary 5.8, noting for all k that $\mathbf{\Delta}_{LTI} \subset \mathbf{\Delta}_k$, there exists a subsequence Δ_{k_n} and a $\Delta \in \mathbf{\Delta}_{LTI}$, $\|\Delta\| \le 1$, so that $\|E_{k_n}\Delta_{k_n} - E\Delta\| \overset{n\to\infty}{\longrightarrow} 0$. Proposition 2.2 states that $\operatorname{rad}(\cdot)$ is continuous at the compact operator $E\Delta$. Hence $\operatorname{rad}(E_{k_n}\Delta_{k_n}) \overset{n\to\infty}{\longrightarrow} \operatorname{rad}(E\Delta)$. From (5.15) we obtain

$$\mu_{\mathbf{\Delta}_{LTI}}(E_{k_n}) \overset{n\to\infty}{\longrightarrow} \operatorname{rad}(E\Delta) \le \mu_{\mathbf{\Delta}_{LTI}}(E).$$

This contradicts (5.14), and therefore $\mu_{\mathbf{\Delta}_{LTI}}(\cdot)$ is necessarily upper semicontinuous.

To demonstrate lower semicontinuity, choose $\epsilon > 0$ and $\Delta \in \mathcal{U}\mathbf{\Delta}_{LTI}$ so that

$$\mu_{\mathbf{\Delta}_{LTI}}(E) < \operatorname{rad}(E\Delta) + \epsilon.$$

Let E_k be any sequence of operators tending to E as $k \to \infty$. Then again by continuity of the spectral radius at a compact operator, we have that $\operatorname{rad}(E_k\Delta) \overset{k\to\infty}{\longrightarrow} \operatorname{rad}(E\Delta)$. This means that for k sufficiently large

$$\mu_{\mathbf{\Delta}_{LTI}}(E) < \operatorname{rad}(E_k\Delta) + \epsilon \le \mu_{\mathbf{\Delta}_{LTI}}(E_k) + \epsilon,$$

which directly implies that $\mu_{\mathbf{\Delta}_{LTI}}(\cdot)$ is lower semicontinuous. ■

The next theorem provides an important continuity result. Note that although $M(z)$ is *not* continuous on $\partial\mathbb{D}$, the function $\mu_{\mathbf{\Delta}_{LTI}}(M(z))$ does possess continuity.

Theorem 5.11 *The structured singular value function $\mu_{\mathbf{\Delta}_{LTI}}(M(z))$ is continuous on $\partial\mathbb{D}$.*

Proof By Proposition 4.19 we know that $M(e^{j\omega})$ is continuous on $(-\pi, \pi]$. Now, at each $\omega_0 \in (-\pi, \pi]$ the operator $M(e^{j\omega_0})$ is compact. Therefore by Proposition 5.10 the function $\mu_{\Delta_{LTI}}(M(e^{j\omega}))$ is continuous on $(-\pi, \pi]$.

To complete the proof we must show that the limit of $\mu_{\Delta_{LTI}}(M(e^{j\omega}))$, as ω tends to $-\pi$ from above, is $\mu_{\Delta_{LTI}}(M(-1))$. Let ω_k be any sequence in $[-\pi, \pi]$ that tends to $-\pi$. Then by Proposition 4.19

$$M(e^{j\omega_k}) \overset{k\to\infty}{\longrightarrow} XM(-1)X^*.$$

The operator $M(-1)$ is compact. Therefore using Proposition 2.2 we have that $XM(-1)X^*$ is compact. So, invoking Proposition 5.10

$$\mu_{\Delta_{LTI}}(M(e^{j\omega_k})) \overset{k\to\infty}{\longrightarrow} \mu_{\Delta_{LTI}}(XM(-1)X^*),$$

where X is defined in (4.16). From Lemma 4.20,

$$\text{RHS} = \mu_{\Delta_{LTI}}(M(-1)). \qquad \blacksquare$$

We require the next lemma to prove the last theorem of this subsection.

Lemma 5.12 *Suppose the sequence ω_k tends to $\omega_0 \in [-\pi, \pi]$. Then*

$$\mu_{\Delta_k}(M_k(e^{j\omega_k})) \overset{k\to\infty}{\longrightarrow} \mu_{\Delta_{LTI}}(M(e^{j\omega_0})).$$

Proof It is sufficient to prove the claim for a subsequence of ω_k: if the result were not true, there would exist a sequence for which no subsequence would satisfy the claim.

Fix $\omega_0 \in [-\pi, \pi]$ and recalling the definition of X in (4.16), set

$$M' := \begin{cases} M(e^{j\omega_0}) & \text{if } \omega_0 \in (-\pi, \pi] \\ XM(-1)X^* & \text{if } \omega_0 = -\pi \end{cases},$$

and note that $\lim_{k\to\infty} M(e^{j\omega_k}) = M'$.

Define $\Delta_k \in \mathcal{UL}(\ell_2)$ so that $\Upsilon_k \Delta_k \Upsilon_k^* \in \mathbf{\Delta}_k$ and satisfy

$$\text{rad}(M(e^{j\omega_k})\Delta_k) \le \mu_{\Delta_k}(M_k(e^{j\omega_k})) \le \text{rad}(M(e^{j\omega_k})\Delta_k) + 2^{-k}.$$

Then by Corollary 5.8 there is a subsequence Δ_{k_n} and a $\Delta \in \mathbf{\Delta}_{LTI}$ satisfying $\|\Delta\| = 1$ and $\lim_{n\to\infty} \|M(e^{j\omega_{k_n}})\Delta_{k_n} - M'\Delta\| = 0$. By continuity of the spectral radius for a compact operator, as in the proof of Theorem 5.9, this implies that $\mu_{\Delta_{k_n}}(M_k(e^{j\omega_{k_n}})) \stackrel{n\to\infty}{\longrightarrow} \mu_{\Delta_{LTI}}(M')$. This now proves the claim by our initial comment. ■

The following theorem provides upper and lower bounds on the stability radius of the sampled-data system. A consequence of the result is that the bounds $\mu_{\Delta_n}(M_n(e^{j\omega}))$ and $\mu_{\underline{\Delta}_n}(\underline{M}_n(e^{j\omega}))$ converge uniformly to $\mu_{\Delta_{LTI}}(M(e^{j\omega}))$ on $(-\pi, \pi]$.

Theorem 5.13 *For every $\epsilon > 0$ there exists n sufficiently large so that*

$$\sup_{\omega\in(-\pi,\pi]} \mu_{\Delta_{LTI}}(M(e^{j\omega})) \leq \sup_{\omega\in(-\pi,\pi]} \mu_{\Delta_n}(M_n(e^{j\omega})) \leq \sup_{\omega\in(-\pi,\pi]} \mu_{\underline{\Delta}_n}(\underline{M}_n(e^{j\omega}))+\epsilon$$

Proof The LHS inequality is a direct consequence of Lemma 5.5. To show the RHS inequality, select a sequence $\omega_n \in [-\pi, \pi]$ so that

$$\tfrac{\epsilon}{3} + \mu_{\Delta_n}(M_n(e^{j\omega_n})) > \sup_{\omega\in(-\pi,\pi]} \mu_{\Delta_n}(M_n(e^{j\omega})).$$

By compactness of $[-\pi, \pi]$, without loss of generality, we may assume that w_n tends to some limit $\omega_0 \in [-\pi, \pi]$. Lemma 5.12 implies, for n large enough, that

$$\tfrac{\epsilon}{3} + \mu_{\Delta_{LTI}}(M(e^{j\omega_0})) > \mu_{\Delta_n}(M(e^{j\omega_n})).$$

The lower bound $\mu_{\underline{\Delta}_n}(\underline{M}_n(e^{j\omega_0}))$ tends to $\mu_{\Delta_{LTI}}(M(e^{j\omega_0}))$ by Theorem 5.1, and therefore for n sufficiently large

$$\tfrac{\epsilon}{3} + \sup_{\omega\in[-\pi,\pi]} \mu_{\underline{\Delta}_n}(\underline{M}_n(e^{j\omega})) > \mu_{\Delta_{LTI}}(M(e^{j\omega_0})).$$

The RHS inequality in the theorem follows directly from the preceding three. ■

We have shown that $\mu_{\Delta_n}(M_n(e^{j\omega_0}))$ is an upper bound for $\mu_{\Delta_{LTI}}(M(e^{j\omega_0}))$ which has the desirable property of uniform monotonic convergence on the interval $(-\pi, \pi]$. We next concentrate on the computation of $\mu_{\Delta_n}(M_n)$, and will demonstrate that it can be obtained to any desired degree of accuracy in terms of a finite dimensional problem.

5.2.2 Characterization in Finite Dimensions

This subsection is devoted to developing a method to compute $\mu_{\Delta_n}(M_n)$. We will show that determining whether $\mu_{\Delta_n}(M_n) \leq r$ is tantamount to determining whether the structured singular value of a finite matrix is bounded by r. Throughout, without loss of generality, we will set $r = 1$ which can always be accomplished by scaling M_n.

To begin, we have the following lemma which states that $\mu_{\Delta_n}(M_n) \leq 1$ can be written as a set of norm constraints.

Lemma 5.14 *Suppose that $W \in \mathfrak{L}(\ell_2)$, $n > 0$, and define $W_n := \Upsilon_n^* W \Upsilon_n$. Given $\mu_{\underline{\Delta}_n}(W_n^{11}) \leq 1$, then $\mu_{\Delta_n}(W_n) \leq 1$ if and only if for each Δ^1 in the open unit ball of $\mathcal{U}\boldsymbol{\Delta}_n$*

$$\|\mathcal{F}_u(W_n, \Delta^1)\| \leq 1 \quad \textit{is satisfied.}$$

The matrix version of this result is originally proved in [47].

Proof Start with the following observation: for any Δ^1 in the open unit ball $\mathcal{U}\underline{\boldsymbol{\Delta}}_n$ we know by hypothesis that $(I - W_n^{11}\Delta^1)^{-1}$ exists and therefore $\mathcal{F}_u(W_n, \Delta^1)$ is defined. For every Δ^2 in $\mathfrak{L}(\ell_2)$ we have

$$\begin{bmatrix} I - W_n^{11}\Delta^1 & -W_n^{12}\Delta^2 \\ -W_n^{21}\Delta^1 & I - W_n^{22}\Delta^2 \end{bmatrix} \cdot \begin{bmatrix} (I - W_n^{11}\Delta^1)^{-1} W_n^{12}\Delta^2 & I \\ I & 0 \end{bmatrix}$$
$$= \begin{bmatrix} 0 & I - W_n^{11}\Delta^1 \\ I - \mathcal{F}_u(W_n, \Delta^1)\Delta^2 & -W_{21}\Delta_1 \end{bmatrix}.$$

Hence, for every $\Delta^1 \in \mathcal{U}\underline{\boldsymbol{\Delta}}_n$ and $\Delta^2 \in \mathfrak{L}(\ell_2)$ we have that $I - \mathcal{F}_u(W_n, \Delta^1)\Delta^2$ is singular if and only if $I - W_n\Delta$ is singular, where $\Delta = \text{diag}(\Delta^1, \Delta^2)$.
(If): Suppose that $\mu_{\Delta_n}(W_n) > 1$. Then there exists Δ in the open unit ball $\mathcal{U}_{\Delta_n}$ so that $I - W_n\Delta$ is singular. By the above discussion this implies $I - \mathcal{F}_u(W_n, \Delta^1)\Delta^2$ is singular, and therefore $\|\mathcal{F}_u(W_n, \Delta^1)\| > 1$ since $\|\Delta^2\| < 1$.
(Only if): Assume for some $\Delta^1 \in \mathcal{U}\underline{\boldsymbol{\Delta}}_n$ that $\|\mathcal{F}_u(W_n, \Delta^1)\| > 1$. Then there exists Δ^2 in the open unit ball of $\mathfrak{L}(\ell_2)$ so that $I - \mathcal{F}_u(W_n, \Delta^1)\Delta^2$ is singular.

■

We now examine the exact representation of M in detail. Recall from (4.12) that the frequency response $M(e^{j\omega_0})$ has the form

$$M(e^{j\omega}) = \tilde{C}e^{j\omega}(I - A_d e^{j\omega})^{-1}\tilde{B} + \tilde{D},$$

where ω_0 is fixed in $[-\pi, \pi]$. The operators are taken from Appendix A. The operator $A_d : \mathbb{C}^{\tilde{n}} \to \mathbb{C}^{\tilde{n}}$ and is given by the matrix

$$A_d := \begin{bmatrix} e^{Ah} + \int_0^h e^{A(h-\eta)}\, d\eta B_2 D_{K_d} C_2 & \int_0^h e^{A(h-\eta)}\, d\eta B_2 C_{K_d} \\ B_{K_d} C_2 & A_{K_d} \end{bmatrix}, \tag{5.16}$$

and

$$\tilde{B} : \ell_2 \to \mathbb{C}^{\tilde{n}} \quad (\tilde{B})_k := h^{-\frac{1}{2}} \begin{bmatrix} I \\ 0 \end{bmatrix} e^{Ah} \int_0^h e^{(j\theta_k - A)\tau}\, d\tau B_1$$

$$\tilde{C} : \mathbb{C}^{\tilde{n}} \to \ell_2 \quad (\tilde{C})_k := h^{-\frac{1}{2}} [C_1 \;\; D_{12}] \int_0^h \exp(\begin{bmatrix} A - j\theta_k & B_2 \\ 0 & -j\theta_k \end{bmatrix} \tau)\, d\tau \cdot \begin{bmatrix} I & 0 \\ D_{K_d} C_2 & C_{K_d} \end{bmatrix}$$

$$\tilde{D} : \ell_2 \to \ell_2 \quad (\tilde{D})_{lk} := h^{-1} C_1 \int_0^h e^{(A - j\theta_l)\tau} \int_0^\tau e^{(j\theta_k - A)\eta}\, d\eta\, d\tau B_1,$$

where

$$\theta_k := \frac{2\pi\nu_k - \omega_0}{h} \tag{5.17}$$

and ν_k is the sequence $\{0, 1, -1, 2, -2, \ldots\}$. Of these operators only $\tilde{D}$ is infinite rank. We proceed to the the following decomposition.

Lemma 5.15 *Suppose that k and l are distinct non-negative integers and that $j\theta_l$ is not an eigenvalue of A. Then*

$$\begin{aligned} (\tilde{D})_{lk} &= (\acute{C})_l \cdot (\acute{B})_k \\ (\tilde{D})_{ll} &= (\acute{C})_l \cdot (\acute{B})_l + (\hat{G}_{11})_l, \end{aligned}$$

where the constituent matrices are given by

$$\begin{aligned} (\acute{C})_l &:= (-h^{-\frac{1}{2}} C_1 \int_0^h e^{(A - j\theta_l)\tau}\, d\tau)(e^{j\omega_0}(I - e^{j\omega_0} e^{Ah})^{-1} \\ &= -h^{-\frac{1}{2}} C_1 (Ij\theta_l - A)^{-1} e^{j\omega_0} \\ (\acute{B})_k &:= (h^{-\frac{1}{2}} e^{Ah} \int_0^h e^{(j\theta_k - A)\tau}\, d\tau B_1) \\ (\hat{G}_{11})_l &:= C_1 (Ij\theta_l - A)^{-1} B_1. \end{aligned}$$

These expressions are routine to verify from (5.16).

In the sequel we have the following *standing* assumption on $n > 0$.

Condition 5.16 *The positive integer $n > 0$ is sufficiently large so that $\bar{\sigma}(C_1(Ij\omega - A)^{-1}B_1) < 1$ is satisfied for all $\omega \notin \cup_{k=0}^{n}[\frac{\pi(2\nu_k-1)}{h}, \frac{\pi(2\nu_k+1)}{h}]$.*

This condition guarantees that $\bar{\sigma}((\hat{G}_{11})_k) < 1$ when $k > n$, and is necessary for our approach to work in general. It is easily checked.

For the moment we fix n and turn our attention to the lifted operator M_n, which by definition is given by

$$M_n = \Upsilon_n M \Upsilon_n^* =: \begin{bmatrix} \tilde{C}_1 \\ \tilde{C}_2 \end{bmatrix} e^{j\omega_0}(I - e^{j\omega_0}A_d)^{-1}[\tilde{B}_1 \;\; \tilde{B}_2] + \begin{bmatrix} \hat{D}_{11} & \hat{D}_{12} \\ \hat{D}_{21} & \hat{D}_{22} \end{bmatrix},$$

where the partitioning of these operators follows from the definition of Υ_n. The operators $\hat{D}_{lk}$ are just the partitioned map $\tilde{D}$. Using Lemma 5.15 we rewrite M_n in the following form:

$$M_n = \begin{bmatrix} \tilde{C}_1 & 0 \\ \tilde{C}_2 & \acute{C}_2 \end{bmatrix} e^{j\omega_0}(I - e^{j\omega_0}\begin{bmatrix} A_d & 0 \\ 0 & 0 \end{bmatrix})^{-1}\begin{bmatrix} \tilde{B}_1 & \tilde{B}_2 \\ 0 & \acute{B}_2 \end{bmatrix} + \begin{bmatrix} \hat{D}_{11} & \hat{D}_{12} \\ \hat{D}_{21} & \hat{G}_{112} \end{bmatrix}, \tag{5.18}$$

where

$$\acute{C}_2 := \begin{bmatrix} (\acute{C})_{n+1} \\ (\acute{C})_{n+2} \\ \vdots \end{bmatrix}; \quad \acute{B}_2 := [(\acute{B})_{n+1} \;\; (\acute{B})_{n+2} \;\; \ldots]$$

$$\hat{G}_{112} := \operatorname{diag}((\hat{G}_{11})_{n+1}, (\hat{G}_{11})_{n+2}, \ldots).$$

From (5.18) we group the operators and adopt the notation

$$M_n =: \left[\begin{array}{c|cc} \hat{A} & \hat{B}_1 & \hat{B}_2 \\ \hline \hat{C}_1 & \hat{D}_{11} & \hat{D}_{12} \\ \hat{C}_2 & \hat{D}_{21} & \hat{G}_{112} \end{array}\right].$$

The fact that M_n is infinite rank is entirely captured in the block-diagonal operator $\hat{G}_{112}$. Our aim is to convert our setup to an equivalent finite rank problem. We use a well-established technique from operator theory; Redheffer [53, Thm I] provides a basis for the following:

Lemma 5.17 *Suppose that $Q \in \mathfrak{L}(\ell_2)$ is a unitary operator, $R \in \mathfrak{L}(\ell_2)$, $\mathcal{F}_u(Q,R)$ is well-defined, and Q_{12} is surjective. Then $\|R\| \le 1$ if and only if $\|\mathcal{F}_u(Q,R)\| \le 1$.*

Proof The proof is based on Figure 5.2:

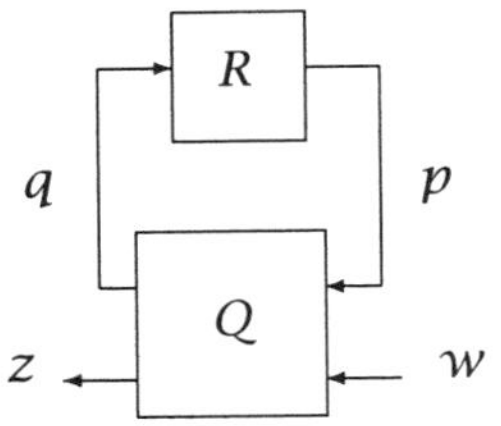

Figure 5.2: Linear Fractional Map

(If): Choose any nonzero q in ℓ_2. Since Q_{12} is surjective there exists $w \in \ell_2$ such that

$$q = Q_{11}Rq + Q_{12}w.$$

Now $\|q\|^2 + \|z\|^2 = \|p\|^2 + \|w\|^2$ because Q is unitary and hence

$$\begin{aligned} \|p\|^2 - \|q\|^2 &= \|z\|^2 - \|w\|^2 \\ &\le (\|\mathcal{F}_u(Q,R)\|^2 - 1)\|w\|^2 \le 0, \end{aligned}$$

because by assumption $\|\mathcal{F}_u(Q,R)\| \le 1$.
(Only if): For any $w \in \ell_2$, since Q is unitary and $\|R\| \le 1$, we have again

$$\begin{aligned} \|z\|^2 - \|w\|^2 &= \|p\|^2 - \|q\|^2 \\ &\le (\|R\|^2 - 1)\|q\|^2 \le 0 \quad \blacksquare \end{aligned}$$

We can apply this result to our configuration as depicted in Figure 5.3; this will give us a finite rank problem.

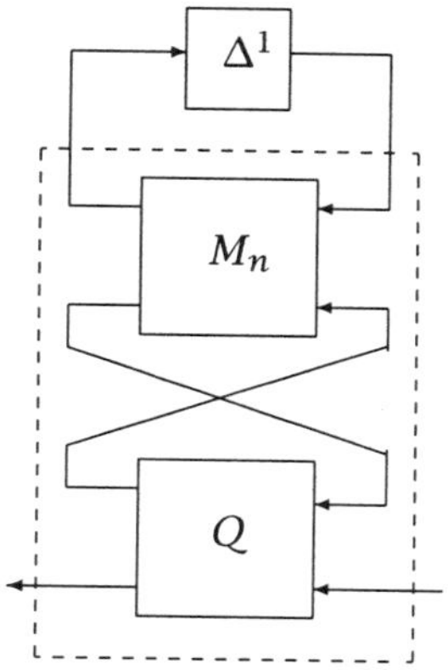

Figure 5.3: Loop-shifted System

Lemma 5.18 *Given $n > 0$ and suppose that*

(i) $\|\widehat{G}_{112}\| < 1$.

(ii) $\mu_{\underline{\Delta}_n}(M_n^{11}) \leq 1$.

(iii) $I - M_n^{22}\widehat{G}_{112}^*$ *is nonsingular.*

Then $\mu_{\Delta_n}(M_n) \leq 1$ *if and only if* $\mu_{\Delta_n}(Q_{M_n}) \leq 1$, *where*

$$Q_{M_n} := \left[\begin{array}{c|cc} \widehat{A} + \widehat{B}_2 T \widehat{C}_2 & \widehat{B}_1 + \widehat{B}_2 T \widehat{D}_{21} & \widehat{B}_2 R^{\frac{1}{2}} \\ \hline \widehat{C}_1 + \widehat{D}_{12} T \widehat{C}_2 & \widehat{D}_{11} + \widehat{D}_{12} T \widehat{D}_{21} & \widehat{D}_{12} R^{\frac{1}{2}} \\ N^{\frac{1}{2}} \widehat{C}_2 & N^{\frac{1}{2}} \widehat{D}_{21} & 0 \end{array}\right]$$

and $R = (I - \widehat{G}_{112}^* \widehat{G}_{112})^{-1}$, $N = (I - \widehat{G}_{112} \widehat{G}_{112}^*)^{-1}$, *and* $T = \widehat{G}_{112}^* N$.

Note that by Lemma 5.5, violating (ii) immediately implies that $\mu_{\underline{\Delta}_n}(M_n) > 1$ since $\mu_{\underline{\Delta}_n}(M_n^{11}) = \mu_{\Delta_n}(\underline{M}_n)$ is a lower bound. Supposition (iii) is also a necessary condition which we will prove in the sequel. Finally, (i) is necessary for our particular decomposition to work; n can always be chosen sufficiently large so that it is satisfied, and it is satisfied when Condition 5.16 holds.

Proof By Lemma 5.14 and (ii) we have that $\mu_{\Delta_n}(M_n) \leq 1$ is equivalent to the condition that for each $\Delta^1 \in \mathcal{U}\underline{\mathbf{\Delta}}_n$,

$$\|\mathcal{F}_u(M_n, \Delta^1)\| \leq 1. \tag{5.19}$$

Since $\|\hat{G}_{112}\| < 1$, the Julia operator

$$Q := \begin{bmatrix} \hat{G}_{112}^* & (I - \hat{G}_{112}^*\hat{G}_{112})^{\frac{1}{2}} \\ (I - \hat{G}_{112}\hat{G}_{112}^*)^{\frac{1}{2}} & -\hat{G}_{112} \end{bmatrix}$$

is well-defined and unitary. By (ii) we know that $(I - M_n^{11}\Delta^1)^{-1}$ exists for all $\Delta^1 \in \mathcal{U}\underline{\mathbf{\Delta}}_n$. Therefore by this and (iii) we have that all the internal signals of Figure 5.3 are uniquely defined, namely the system is well-posed. Hence, $\mathcal{F}_u(Q, \mathcal{F}_u(M_n, \Delta_n^1))$ is defined for each such Δ_n^1.

We now invoke Lemma 5.17 to obtain that (5.19) holds, for all $\Delta^1 \in \mathcal{U}\underline{\mathbf{\Delta}}_n$, if and only if

$$\|\mathcal{F}_u(Q, \mathcal{F}_u(M_n, \Delta^1))\| \leq 1 \text{ is satisfied for each } \Delta^1 \in \mathcal{U}\underline{\mathbf{\Delta}}_n. \tag{5.20}$$

The map Q_{M_n} defined above is exactly the operator inside the dashed box in Figure 5.3. Therefore (5.20) is equivalent to the statement $\|\mathcal{F}_u(Q_{M_n}, \Delta_n^1)\| \leq 1$ for all $\Delta_n^1 \in \mathcal{U}\underline{\mathbf{\Delta}}_n$. Lemma 5.14 states that this is equivalent to $\mu_{\Delta_n}(Q_{M_n}) \leq 1$ provided that $\mu_{\Delta_n}(Q_{M_n}^{11}) \leq 1$. The latter inequality must hold, because as shown, the system is well-posed. It is routine to verify that Q_{M_n} is the argument operator given in the lemma statement. ■

We now show that condition (iii) of the lemma is necessary for the claim to hold.

Lemma 5.19 *Suppose that $\|\hat{G}_{112}\| < 1$. If $\mu_{\Delta_n}(M_n) < 1$, then $I - M_n^{22}\hat{G}_{112}^*$ is nonsingular. Furthermore, the operator $I - M_n^{22}\hat{G}_{112}^*$ is nonsingular if and only if the matrix $I - e^{j\omega_0}(\hat{A} + \hat{B}_2 T\hat{C}_2)$ is nonsingular.*

The lemma states that it is necessary that the "A" matrix in Lemma 5.18 not have an eigenvalue at $e^{-j\omega_0}$. In the next subsection we will derive a state space expression for this matrix.

Proof Suppose that $I - M_n^{22}\hat{G}_{112}^*$ is singular. Then $\Delta_n = \operatorname{diag}(0, \hat{G}_{112}^*)$ is in $\mathcal{U}\boldsymbol{\Delta}_n$, and $I - M_n\Delta_n$ is also singular. Hence, $\mu_{\Delta_n}(M_n) > 1$. The second part of the claim follows routinely by recalling that $M_n^{22} = \hat{C}_2 e^{j\omega_0}(I - e^{j\omega_0}\hat{A})^{-1}\hat{B}_2 + \hat{G}_{112}$. ■

Our final step in this construction is to convert our test on the finite rank operator in Lemma 5.18, to one on a matrix. To do this define the matrices $\overline{B}_1$, $\overline{C}_2$, $\overline{D}_{12}$ and $\overline{D}_{21}$ to be any satisfying

$$\begin{bmatrix} \hat{B}_2 \\ \hat{D}_{12} \end{bmatrix} R\,[\hat{B}_2^* \;\; \hat{D}_{12}^*] =: \begin{bmatrix} \overline{B}_2 \\ \overline{D}_{12} \end{bmatrix} [\overline{B}_2^* \;\; \overline{D}_{12}^*] \tag{5.21}$$

$$\begin{bmatrix} \hat{C}_2^* \\ \hat{D}_{21}^* \end{bmatrix} N\,[\hat{C}_2 \;\; \hat{D}_{21}] =: \begin{bmatrix} \overline{C}_2^* \\ \overline{D}_{21}^* \end{bmatrix} [\overline{C}_2 \;\; \overline{D}_{21}].$$

Such matrices always exist since the LHS expressions both represent positive semi-definite matrices. From Lemma 5.15 it follows that the ranks of these matrices are equal to the ranks of $\hat{B}_2$ and $\hat{C}_2$ respectively; so their ranks are bounded for all n from the definitions of $\hat{B}_2$ and $\hat{C}_2$ in (5.18). Given the LHS, solutions to the RHS can be obtained, for example, by Schur decomposition or Cholesky factorization.

Having obtained such matrices we define the matrix

$$\overline{M}_n := \left[\begin{array}{c|cc} \hat{A} + \hat{B}_2 T \hat{C}_2 & \hat{B}_1 + \hat{B}_2 T \hat{D}_{21} & \overline{B}_2 \\ \hline \hat{C}_1 + \hat{D}_{12} T \hat{C}_2 & \hat{D}_{11} + \hat{D}_{12} T \hat{D}_{21} & \overline{D}_{12} \\ \overline{C}_2 & \overline{D}_{21} & 0 \end{array}\right]. \tag{5.22}$$

Associated with this matrix we define the uncertainty set $\bar{\boldsymbol{\Delta}}_n$, define b to be the number of columns of $\overline{B}_2$ and c the number of rows of $\overline{C}_2$, and

$$\bar{\boldsymbol{\Delta}}_n := \{\operatorname{diag}(\Delta^1, \Delta^2) : \Delta^1 \in \underline{\boldsymbol{\Delta}}_n,\ \Delta^2 \in \mathbb{C}^{b\times c}\}.$$

We can now achieve the main objective of this subsection:

Theorem 5.20 *Suppose that conditions (i)–(iii) in Lemma 5.18 are satisfied. Then $\mu_{\Delta_n}(M_n) \le 1$ if and only if $\mu_{\bar{\Delta}_n}(\overline{M}_n) \le 1$.*

The theorem asserts that determining whether our upper bound is less than one is equivalent to determining whether the structured singular value of a matrix is less than one. This provides a means to iteratively obtain $\mu_{\Delta_n}(M_n)$, using for example bisection, and checking the bounds on the structured singular value $\mu_{\bar{\Delta}_n}(\overline{M}_n)$ at each step.

Proof To begin we observe that both $R^{\frac{1}{2}}[\hat{B}_2^* \ \ \hat{D}_{12}^*]$ and $N^{\frac{1}{2}}[\hat{C}_2 \ \ \hat{D}_{21}]$ are finite rank operators and therefore their images are isomorphic to Euclidean space: from (5.21) we see that

$$\begin{aligned} \langle R^{\frac{1}{2}}[\hat{B}_2^* \ \ \hat{D}_{12}^*]v, R^{\frac{1}{2}}[\hat{B}_2^* \ \ \hat{D}_{12}^*]u \rangle_2 &= \langle [\bar{B}_2^* \ \ \bar{D}_{12}^*]v, [\bar{B}_2^* \ \ \bar{D}_{12}^*]u \rangle \\ \langle N^{\frac{1}{2}}[\hat{C}_2 \ \ \hat{D}_{21}]r, N^{\frac{1}{2}}[\hat{C}_2 \ \ \hat{D}_{21}]p \rangle_2 &= \langle [\bar{C}_2 \ \ \bar{D}_{21}]r, [\bar{C}_2 \ \ \bar{D}_{21}]p \rangle, \end{aligned} \tag{5.23}$$

hold for all $u, v \in \mathbb{C}^{\hat{n}+pn}$ and $r, p \in \mathbb{C}^{\hat{n}+pn}$. Define $Y : \ell_2 \to \mathbb{C}^b$ to be the unique operator that satisfies

$$YR^{\frac{1}{2}}[\hat{B}_2^* \ \ \hat{D}_{12}^*] = [\bar{B}_2^* \ \ \bar{D}_{12}^*],$$

and $Yx = 0$ for all x in the orthogonal complement of the image of $R^{\frac{1}{2}}[\hat{B}_2^* \ \ \hat{D}_{12}^*]$. By (5.23) it follows that $\|Y\| \le 1$. Similarly define $V : \ell_2 \to \mathbb{C}^c$ to be the the map satisfying $VN^{\frac{1}{2}}[\hat{C}_2 \ \ \hat{D}_{21}] = [\bar{C}_2 \ \ \bar{D}_{21}]$ that is zero on the orthogonal complement of the image of $N^{\frac{1}{2}}[\hat{C}_2 \ \ \hat{D}_{21}]$. Define

$$J_V := \begin{bmatrix} I & 0 \\ 0 & V \end{bmatrix}; \quad J_Y := \begin{bmatrix} I & 0 \\ 0 & Y \end{bmatrix}.$$

We can now easily demonstrate the theorem.
(If): By Lemma 5.18 it is sufficient to show that $\mu_{\Delta_n}(Q_{M_n}) \le \mu_{\bar{\Delta}_n}(\overline{M}_n)$. Suppose there exists $\Delta_n \in \mathbf{\Delta}_n$ so that $I - Q_{M_n}\Delta_n$ is singular. It follows that

$$J_V J_V^* - (J_V M_n J_Y^*)(J_Y \Delta_n J_V^*)$$

is singular, since it is straightforward to show that $Q_{M_n} = Q_{M_n} J_Y^* J_Y$. From the above definitions it is routine to verify that $J_V J_V^* = I$ and

$J_V Q_{M_n} J_Y^* = \overline{M}_n$. i.e. $I - \overline{M}_n (J_Y \Delta_n J_V^*)$ is singular. Now, $(J_Y \Delta_n J_V^*) \in \bar{\Delta}_n$ and $\|(J_Y \Delta_n J_V^*)\| \leq \|\Delta_n\|$ because J_Y and J_V are contractive; hence $\mu_{\Delta_n}(Q_{M_n}) \leq \mu_{\bar{\Delta}_n}(\overline{M}_n)$.
(Only if): This follows by starting with $I - \overline{M}_n \bar{\Delta}$ singular and reversing the above argument. ■

Remark 5.21 *The decomposition used here is much simpler when the transfer function $C_1(Is - A)^{-1} B_1$ has no imaginary axis poles. Refer to Appendix C for the details.*

We end this section with a corollary to the theorem in which $\overline{M}_n$ depends explicitly on frequency.

Corollary 5.22 *Given that n satisfies Assumption 5.16 and $r > 0$. Figure 4.1 has robust stability to $r\mathcal{U}\mathfrak{X}_{LTI}$ if*

$$r \sup_{\omega \in (-\pi, \pi]} \mu_{\bar{\Delta}_n}(\overline{M}_n(e^{j\omega})) \leq 1.$$

5.2.3 Evaluating $\overline{M}_n$

In order to apply the result of Theorem 5.20 we require exact expressions for the matrices which constitute $\overline{M}_n$ in (5.22). The matrices $\hat{A}$, $\hat{B}_1$, $\hat{C}_1$, and $\hat{D}_{11}$ can be obtained from their definition in (5.18), but the remaining entries in $\overline{M}_n$ are stated in terms of the products of operators. To illustrate the technique of evaluating these terms we proceed to derive the formula for $\hat{B}_2 T \hat{C}_2$ explicitly.

To start we look at the factors of $\hat{B}_2 T \hat{C}_2$: the operator $\hat{B}_2$ maps ℓ_2 to $\mathbb{C}^{\hat{n}}$ and can therefore be viewed as an infinite matrix of the form $[(\hat{B})_{n+1} \ (\hat{B})_{n+2} \ \ldots]$. Each block entry is given by

$$(\hat{B})_k := C_B (Ij\theta_k - A)^{-1} B_1,$$

which can be verified from (5.18); the matrix C_B is independent of the k and is provided in Appendix A. Similarly, $\hat{C}_2$ is a matrix with an infinite number of rows:

$$\widehat{C} =: \begin{bmatrix} (\widehat{C})_{n+1} \\ (\widehat{C})_{n+2} \\ \vdots \end{bmatrix}; \quad (\widehat{C})_k := [C_1 \ \ 0](Ij\theta_k - \begin{bmatrix} A & B_2 \\ 0 & 0 \end{bmatrix})^{-1}B_C,$$

where B_C is a matrix that can be found in Appendix A. From its definition in Lemma 5.18, we have $T = \widehat{G}_{112}^*(I - \widehat{G}_{112}\widehat{G}_{112}^*)^{-1}$. Now $\widehat{G}_{112}$ is diagonal and therefore so is T. We have $T =: \text{diag}(T_{n+1}, T_{n+2}, \ldots)$ where $(T)_k := (\widehat{G}_{11}^*)_k(I - (\widehat{G}_{11})_k(\widehat{G}_{11}^*)_k)^{-1}$. From the equation in Lemma 5.15 we can show that

$$(T)_k := [0 \ \ -B_1^*](Ij\theta_k - \begin{bmatrix} A & B_1B_1^* \\ -C_1C_1^* & -A^* \end{bmatrix})^{-1}\begin{bmatrix} 0 \\ C_1^* \end{bmatrix}.$$

With these representations we arrive at

$$\widehat{B}_2 T \widehat{C}_2 = \sum_{k=n+1}^{\infty} (\widehat{B})_k (T)_k (\widehat{C})_k.$$

Each term in the summation has the form of the product of three continuous-time transfer functions evaluated at $j\theta_k$, and by standard state-space manipulations we get

$$(\widehat{B})_k(T)_k(\widehat{C})_k = C_B[I \ \ 0 \ \ 0 \ \ 0](Ij\theta_k - A_W)^{-1}\begin{bmatrix} 0 \\ I \end{bmatrix} B_C, \tag{5.24}$$

where A_W is a matrix independent of ω_0 and k; see Appendix A.

To evaluate the infinite sum we require a result:

Proposition 5.23 *Suppose that $e^{j\omega_0}$ is not an eigenvalue of the square matrix e^{Eh}. Then*

$$\sum_{k=0}^{\infty}(Ij\theta_k - E)^{-1} = -j\frac{h}{2}\cot(\,(jEh + \omega_0 I)/2\,),$$

where the sequence $\theta_k := (2\pi\nu_k - \omega_0)h^{-1}$.

This formula can also be found in [39]; we provide an independent and elementary proof below.

Proof Consider the function $E(t) := h(I - e^{Eh}e^{j\omega_0})^{-1}e^{(E+j\omega_0/h)t}$ on the interval $[0, h]$: its Fourier series is

$$E(t) \sim \sum_{k=0}^{\infty}(Ij\theta_k - E)^{-1}e^{\frac{2\pi}{h}\nu_k t},$$

where again ν_k is the sequence $\{0, 1, -1, 2, -2, \ldots\}$. Because $E(t)$ is of bounded variation and continuous on the interval, it is well-known from classical Fourier analysis that, at $t = 0$, the series sums to the average of the end points $(E(0)+E(h))/2$ (see e.g. [76]), which is just $-j\frac{h}{2}\cot((jEh+\omega_0 I)/2)$. ■

Related to this series we define a function.

Definition 5.24 *Given a square matrix E and $n > 0$, and the fact that E has no eigenvalues equal to $j\theta_k$ for $k > n$, define the function*

$$F_n(E) := \sum_{k=n+1}^{\infty}(Ij\theta_k - E)^{-1}.$$

The function can be evaluated, using the proposition, by

$$F_n(E) = -j\frac{h}{2}\cot((jEh + \omega_0 I)/2) - \sum_{k=0}^{n}(Ij\theta_k - E)^{-1},$$

except when $e^{j\omega_0}$ is an eigenvalue of E. This occurs at most at dim(E) values of ω_0 on $(-\pi, \pi]$.

Turning back to (5.24) it is easy to see that A_W has eigenvalues corresponding to those of $\begin{bmatrix} A & B_1B_1^* \\ -C_1C_1^* & -A^* \end{bmatrix}$ and $\begin{bmatrix} A & B_2 \\ 0 & 0 \end{bmatrix}$. By Condition 5.16 we know that $\bar{\sigma}((\widehat{G}_{11})_k) < 1$ for $k > n$. Therefore, neither of these matrices have eigenvalues equal to $j\theta_k$ for $k > n$ and $F_n(A_W)$ is defined. Hence, we get immediately from (5.24) that

$$\widehat{B}_2T\widehat{C}_2 = C_B[I \ \ 0 \ \ 0 \ \ 0]F_n(A_W)\begin{bmatrix} 0 \\ I \end{bmatrix}B_C.$$

The remaining component matrices for $\overline{M}_n$ can be found in Appendix A. All have been obtained using manipulations similar to those above.

Remark 5.25 *If we replace the sequence ν_k, defined in equation (5.17), by some other reordering of the integers, all the results so far hold, except for Proposition 5.23 which does not hold in general. However, more care must also be taken in defining the function F_n above.*

We have now completed the development of the upper bound. The next section presents a simple algorithm to compute bounds for $\mu_{\Delta_{LTI}}(M(e^{j\omega_0}))$ in terms of the results of this chapter.

5.3 An Algorithm

We outline an algorithm to compute bounds for $\mu_{\Delta_{LTI}}(M)$ at a frequency point ω_0 of interest. The bounds developed in the previous sections are in terms of the structured singular values of matrices. To date methods exist for computing upper and lower bounds on the structured singular value of general matrices; a synopsis of the current results in this area is provided in [47]. The specific bounds used are those of Proposition 2.15.

These methods can therefore be used to compute bounds on $\mu_{\underline{\Delta}_n}(\underline{M}_n)$ and $\mu_{\bar{\Delta}_n}(\overline{M}_n)$ which are related to our sampled-data problem. In the algorithm presented below $\underline{L}_n$ and $\underline{U}_n$ signify the lower and upper bounds, respectively, for $\mu_{\underline{\Delta}_n}(\underline{M}_n)$. Similarly $\overline{L}_n$ and $\overline{U}_n$ are lower and upper bounds for $\mu_{\Delta_n}(M_n)$, which must be obtained iteratively by calculating bounds for $\mu_{\bar{\Delta}_n}(\overline{M}_n)$ and invoking Theorem 5.20.

Algorithm

1. Compute bounds $\underline{L}_0 \le \mu_{\underline{\Delta}_0}(\underline{M}_0) \le \underline{U}_0$.

2. Choose n sufficiently large so that $\underline{L}_0^{-1} C_1 (Ij\omega - A)^{-1} B_1$ satisfies Condition 5.16.

3. Compute bounds $\underline{L}_n \le \mu_{\underline{\Delta}_n}(\underline{M}_n) \le \underline{U}_n$.

4. Use Theorem 5.20 to compute bounds $\overline{L}_n \le \mu_{\Delta_n}(M_n) \le \overline{U}_n$, using bisection and bounds for $\mu_{\bar{\Delta}_n}(\overline{M}_n)$.

5. If $\overline{L}_n$ and $\underline{U}_n$ are not sufficiently close, increment n and repeat (3–4). Otherwise $\underline{L}_n \leq \mu_{\Delta_{LTI}}(M) \leq \overline{U}_n$ holds.

We now make some general remarks about this algorithm. The accuracy to which we can compute $\mu_{\Delta_{LTI}}(M)$ is clearly dependent on our ability to get good bounds for $\mu_{\underline{\Delta}_n}(\underline{M}_n)$ and $\mu_{\bar{\Delta}_n}(\overline{M}_n)$. There are no theoretically based guarantees that these bounds will be close; however the numerical experience of researchers in the area of structured singular value computation indicates that the bounds are usually close. See for example [4]. We also point out that the matrices $\underline{M}_n$ and $\overline{M}_n$ have considerable structure. Therefore methods that exploit this structure may provide advantage. This is a potential topic for future research.

Recall that computing $\mu_{\underline{\Delta}_n}(\underline{M}_n)$ corresponds to computing the structured singular value of the matrix $\underline{M}_n$ with a perturbation of $n+1$ structured blocks. Calculating $\mu_{\bar{\Delta}_n}(\overline{M}_n)$ involves obtaining a structured singular value with an uncertainty structure in which there are $n+2$ structured blocks. The numerical complexity of computing the structured singular value increases with the number of blocks in the uncertainty structure. Therefore, as n increases, so does the computational cost associated with computing these upper and lower bounds for $\mu_{\Delta_{LTI}}(M)$. At the same time, increasing n monotonically improves the tightness of the upper and lower bounds. Hence, a tradeoff between the accuracy of the bounds and the computational effort required to obtain them is established.

To complete this section we observe that the result of Theorem 4.15 can be combined with $\mathcal{H}_\infty$-optimal sampled-data design, in the style of μ-synthesis described in [21], as a basis to obtain controllers for our robust stabilization problem. The idea is as follows: for a fixed $n > 0$ and $\omega_l \in (-\pi, +\pi]$, define $D \in \mathcal{D}_{\underline{\Delta}_n}$ to be a nonsingular complex matrix which commutes with every element of $\underline{\Delta}_n$, as described in Section 2.6, and is chosen to minimize $\bar{\sigma}(D^{-1}\underline{M}_n(e^{j\omega_l})D)$. Such a 'D-scale' always satisfies $\bar{\sigma}(D^{-1}\underline{M}_n(e^{j\omega_l})D) \geq \mu_{\underline{\Delta}_n}(\underline{M}_n(e^{j\omega_l}))$, from Proposition 2.15 and has the form $D = \text{diag}(\underline{D}_0, \ldots, \underline{D}_n)$. By analogy with Corollary 4.8, each $\underline{D}_k$ can be associated with a dynamic scaling, $\hat{D}(s) \in \mathcal{A}_{\mathbb{R}}$, which satisfies $\hat{D}(j\frac{2\pi\nu_k-\omega_l}{h}) = \underline{D}_k$. If scalings are computed for a finite number of discrete-time frequencies $\omega_l \in (-\pi, \pi]$, we obtain values of $\hat{D}(j\omega)$ over the continuous-time frequency variable in the range $\cup_{k=0}^{\infty}[\frac{\pi(2\nu_k-1)}{h}, \frac{\pi(2\nu_k+1)}{h}]$. After obtaining a stable rational interpolating function $\hat{D}(s)$, we can use it to augment the plant in the usual way, and then perform an $\mathcal{H}_\infty$-sampled data synthesis on this augmented plant by the methods of [7] and [32] to gain our controller. Here we have only provided the idea for such a synthesis procedure, which could be a future research direction. References [21] and [72] provide a good basis for studying such a practical synthesis approach.

5.4 Example

We use the feedback configuration shown in Figure 5.4 to demonstrate the computations of the robust stability bounds derived. It has been chosen to keep this illustrative example as simple as possible. The setup can be described by the more general arrangement of Figure 3.1 and is analyzed for various values of a.

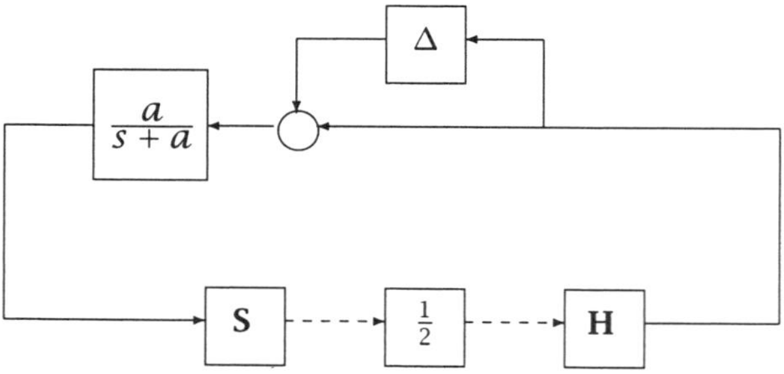

Figure 5.4: Additive Perturbation

Varying the cutoff parameter a of the filter, with the sampling period fixed at h =ln2 we compute the stability radius of the system; we also compare this robustness condition with the $\mathcal{L}_2 \to \mathcal{L}_2$ induced norm small gain theorem. For comparison, we include the $\mathcal{L}_\infty \to \mathcal{L}_\infty$ induced norm small gain theorem results, [37][59].

a	Thm 4.15	$\mathcal{L}_2$ SGT	$\mathcal{L}_\infty$ SGT
2	0.99	0.93	1.00
100	0.78	0.18	1.00
1000	0.6	0.05	1.00

The first column in the table gives the value a chosen for the filter. Then each of the following columns provides the value of the largest ball of perturbations to which that particular test guarantees robustness. The calculations of the $\mathcal{L}_2 \to \mathcal{L}_2$ and $\mathcal{L}_\infty \to \mathcal{L}_\infty$ induced norms were done using the methods of [8] and [24], respectively.

From the table we see that as the filter is changed, the $\mathcal{L}_2$ small gain theorem predicts significantly reduced robustness: the allowable perturbation ball goes from 0.93 down to 0.05. However for LTI perturbations, using Theorem 4.15 and Theorem 5.20, we find that robustness is guaranteed for larger balls, ranging from 0.99 down to 0.6. Hence, for this example the small gain theorem is quite conservative for large values of a when applied to LTI perturbations. We point out that at these values the filter bandwidth is much greater than the sampling rate of $1/\ln 2$, and hence does not follow the usual engineering rule-of-thumb.

In Figure 5.5 we plot the family of bounds obtained from Theorem 4.15 and Theorem 5.20 when $a = 2$ and the sampling period $h = \ln 2$. Here, the increasing sequence of lower bound functions $\mu_{\underline{\Delta}_n}(\underline{M}_n(e^{j\omega}))$ are shown by the dotted lines for $n = 1, 3, 10$. The upper bound function $\mu_{\Delta_1}(M_1(e^{j\omega}))$ is shown by the solid line. We show the transfer function norm $\|M(e^{jw}\|$ by the dashed line. Both $\mu_{\underline{\Delta}_n}(\underline{M}_n(e^{j\omega}))$ and $\mu_{\Delta_n}(M_n(e^{j\omega}))$ have been calculated exactly; this is possible because these are rank one structured singular value problems. We remark that the lower bound can be evaluated as a scalar sum using Proposition 5.3. From the plot we see that the function $\mu_{\Delta_{LTI}}(M(e^{j\omega}))$ is nearly captured by $\mu_{\underline{\Delta}_{10}}(\underline{M}_{10}(e^{j\omega}))$ and $\mu_{\Delta_1}(M_1(e^{j\omega}))$; it has its maximum of approximately 1 at the origin. Hence, the stability radius of the system with respect to LTI perturbations is also about 1. At $\omega = 0$ the function $\|M(e^{jw}\|$ has its maximum of 1.08 and therefore the

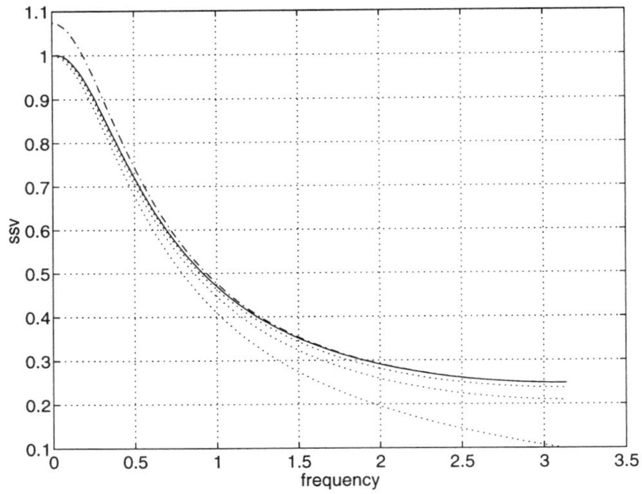

Figure 5.5: Upper and Lower Bounds for $\mu_{\Delta_{LTI}}(M(e^{j\omega}))$ with $a = 2$

small gain condition predicts a stability radius of 0.93.

Figure 5.6 plots the upper and lower bounds when the system parameter $a = 100$ and $h = \ln 2$. We remark that at this value of a the system is undersampled according to traditional engineering heuristics. The lower bounds $\mu_{\underline{\Delta}_n}(\underline{M}_n(e^{j\omega}))$ are shown for $n = 1, 5, 10, 100$ and the upper bounds $\mu_{\Delta_n}(M_n(e^{j\omega}))$ are given with $n = 1, 3, 10$, again using dotted and solid lines respectively. As with Figure 5.5, the lower bound functions are slower to converge to $\mu_{\Delta_{LTI}}(M(e^{j\omega}))$ than are the upper bounds: $\mu_{\underline{\Delta}_n}(\underline{M}_n(e^{j\omega}))$ continues to vary significantly with n until about 100 blocks are used; in contrast the family $\mu_{\Delta_n}(M_n(e^{j\omega}))$ nearly converged when $n = 10$.

From the convergence of the bounds we see that the maximum of $\mu_{\Delta_{LTI}}(M(e^{j\omega}))$ occurs at $\omega = 0.55$ and is roughly equal to 1.28, giving an LTI stability radius of 0.78. The transfer function norm $\|M(e^{j\omega})\|$ is shown by the dashed line and we see that it has a maximum of 5.8 at the

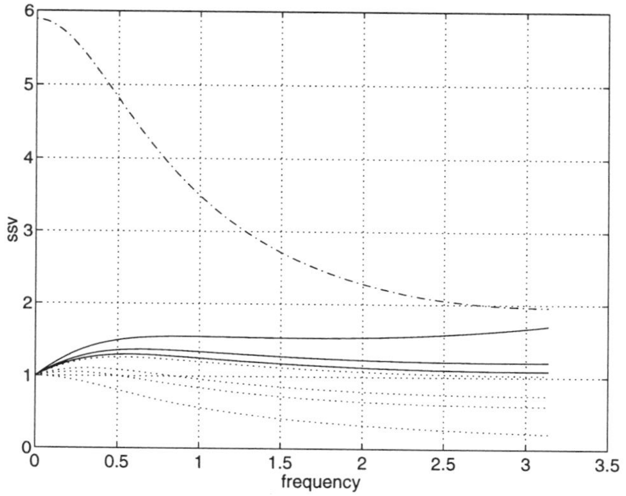

Figure 5.6: Upper and Lower Bounds for $\mu_{\Delta_{LTI}}(M(e^{j\omega}))$ with $a = 100$

origin, yielding a small gain stability radius of 0.18. Note that with $n = 1$, $\mu_{\Delta_1}(\underline{M}_1(e^{j\omega}))$ and $\mu_{\Delta_1}(M_1(e^{j\omega}))$ bound the stability radius to within the interval [0.55, 1].

5.5 Summary

A computational method has been developed to evaluate the stability test of Chapter 4. This scheme provides means by which to compute upper and lower bounds on the structured singular $\mu_{\Delta_{LTI}}(M(e^{j\omega}))$, which can be chosen to be as close as desired. Thus computing bounds for $\mu_{\Delta_{LTI}}(M(e^{j\omega}))$ is reduced to two finite dimensional optimization problems, which grow in complexity as the accuracy of the bounds is monotonically improved.

Also considered in this chapter were the continuity properties of the structured singular value $\mu_{\Delta_{LTI}}(\cdot)$ on the compact operators. These results ensured that the computational scheme developed can be used to uniformly approximate the desired quantity $\mu_{\Delta_{LTI}}(M(e^{j\omega}))$.

Section 5.3 provides an explicit algorithm for evaluating robust stability to LTI perturbations using the developed bounds. It also suggests an idea for a synthesis procedure.

Finally, a simple example is presented that illustrates the various computations of the chapter. The example also shows that there can be a very significant quantitative difference between the stability radius of a sampled-data system to unstructured LTI perturbations and the stability radius predicted by the small gain theorem; this is of note because such disparities do not occur in purely continuous time LTI systems.

At this point we have completed our analysis of structured LTI uncertainty. In the next chapter we investigate robustness to a larger class of structured uncertainty, one that allows more time variation. The LTI robustness conditions we have developed are preferred over the time varying robustness conditions of the next chapter, which are less complex, in cases where engineering motivation justifies the increased computational effort required for LTI robustness analysis.

Chapter 6

Robust Performance

We have considered robust stabilization to LTI perturbations; that is, perturbations which do not vary with time. In this chapter we concentrate on three expanded perturbation classes each permitting a different level of time variation. We focus on robust performance analysis of sampled-data systems which are subject to perturbations that are periodic in the sampling rate of the nominal system, quasi-periodic and arbitrary time-varying.

Expanding the perturbation classes in this way has two main advantages. First, as we will see, it makes the treatment of robust performance simpler because the uncertainty and performance specifications can be considered in a unified manner. The second major advantage is that the computations resulting from the analysis are less complex. Of course, as shown by the example of the last chapter, there can be a significant difference even between LTI and periodic robustness; however in many practical applications it may be clear from the engineering motivation that this is not an important consideration.

6.1 Robust Performance Conditions

Throughout we focus on the robust performance arrangement depicted in Figure 3.1. For convenience we rewrite this configuration, as shown in

Figure 6.1, by setting $\mathbf{M} = \mathcal{F}_l(\mathbf{G}, \mathbf{HK_dS})$. Hence, we have an equivalent representation of Figure 3.4. The key differences in this setup from that of the previous chapter is the presence of the exogenous input w, and the regulated output z. Our goal is to derive exact conditions for robust performance, defined in Definition 3.2, to three new uncertainty classes. As before, the perturbation classes will all be subsets of the structured perturbation class $\mathfrak{X}_s$ defined in Chapter 3.

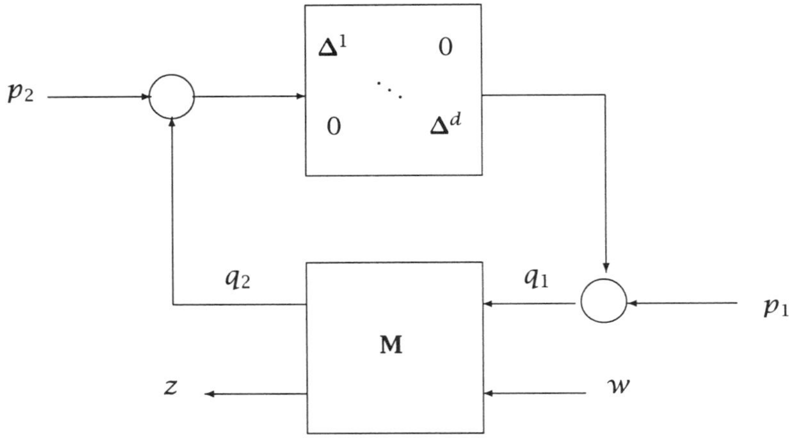

Figure 6.1: Robust Performance Configuration

The section is divided into three subsections: the first deals with robust performance to periodic perturbations, the second with quasi-periodic perturbations and the final subsection provides robustness results on arbitrary time-varying perturbations. To construct the various performance conditions we work within the sampled-data robustness framework developed in Chapters 3 and 4. However, to analyze these new perturbation classes, in contrast to the LTI uncertainty analysis, we do not require the use of the sampled-data frequency response. It will suffice to work over the space of functions $\mathcal{A}$.

6.1.1 Periodic Perturbations

We start by defining the perturbation class of periodic operators to be treated. Recall the following definition

$$\mathfrak{L}_{\mathcal{A}} := \{\mathbf{\Delta} \in \mathfrak{L}(\mathcal{L}_2) : \text{ there exists } \check{\Delta} \in \mathcal{A} \text{ so that } \mathbf{\Delta} = W^{-1}Z^{-1}\Theta_{\check{\Delta}}ZW\},$$

where W is the sampled-data lifting operator defined in (2.4) and Z is the transform defined in (2.7). In words, $\mathfrak{L}_{\mathcal{A}}$ is the subspace of operators that have operator-valued transfer functions in $\mathcal{A}$. All the elements of $\mathfrak{L}_{\mathcal{A}}$ are h-periodic: for each $\mathbf{\Delta} \in \mathfrak{L}_{\mathcal{A}}$ we have that

$$\mathbf{D}_h\mathbf{\Delta} = \mathbf{\Delta}\mathbf{D}_h,$$

where $\mathbf{D}_h$ is the delay by h on $\mathcal{L}_2$. Also recall that this does not include all causal h-periodic operators on $\mathcal{L}_2$; the space of all such operators is isomorphic to a larger subspace of $\mathcal{H}_\infty$. For technical convenience we prefer to work with operators whose transfer functions are in $\mathcal{A}$ because their boundary values on the unit circle are well-defined everywhere.

The specific perturbation class we consider is the set of structured periodically time-varying operators given by

$$\mathfrak{X}_{PTV} := \mathfrak{L}_{\mathcal{A}} \cap \mathfrak{X}_s$$

where $\mathfrak{X}_s$ is the set of spatially structured operators defined in (3.2). Hence, for each $\mathbf{\Delta} \in \mathfrak{X}_{PTV}$ there corresponds a transfer function $\check{\Delta} \in \mathcal{A}$ such that

$$\check{\Delta}(z) := \begin{bmatrix} \check{\Delta}_1(z) & & 0 \\ & \ddots & \\ 0 & & \check{\Delta}_d(z) \end{bmatrix},$$

where each $\check{\Delta}_k \in \mathcal{A}$ and maps $\bar{\mathbb{D}}$ to $\mathfrak{L}(\mathcal{K}_2^{m_k})$. Define the subspace of $\mathfrak{L}(\mathcal{K}_2^m)$ that these spatially structured transfer functions take values in:

$$\boldsymbol{\Delta}_{PTV} := \{\Delta \in \mathfrak{L}(\mathcal{K}_2^m) : \Delta = \operatorname{diag}(\Delta_1, \ldots, \Delta_d)\}. \tag{6.1}$$

That is, an operator $\mathbf{\Delta}$ is in $\mathfrak{X}_{PTV}$ if and only if it has a transfer function $\check{\Delta} \in \mathcal{A}$ that maps $\bar{\mathbb{D}}$ to $\boldsymbol{\Delta}_{PTV}$.

Our main intention is now to characterize robust performance of Figure 6.1 to perturbations in $\mathfrak{X}_{PTV}$. This, we show, can be formulated as a frequency domain structured singular value problem.

First we have the following interpolation result for perturbations in our uncertainty set.

Lemma 6.1 *Suppose $\Delta_0 \in \mathcal{U}\boldsymbol{\Delta}_{PTV}$ and that ω_0 is nonzero and in the interval $(-\pi, \pi)$. There exists a causal operator $\boldsymbol{\Delta} \in \mathcal{U}\mathfrak{X}_{PTV}$, so that its transfer function in $\boldsymbol{\mathcal{A}}$ satisfies*

$$\check{\Delta}(e^{j\omega_0}) = \Delta_0.$$

This result states that, except for at the frequencies zero and π, we can interpolate any value in $\boldsymbol{\Delta}_{PTV}$ with a transfer function in $\boldsymbol{\mathcal{A}}$. Note that in the proof below the perturbation constructed is real.

Proof We set

$$\begin{aligned}
&\check{\Delta}(z) := \\
&z\left(\frac{e^{j\omega_0} - (1+1/\alpha)e^{-j\omega_0}}{e^{j2\omega_0} - 1} \frac{z - e^{-j\omega_0}}{z - (1+1/\alpha)e^{-j\omega_0}} \frac{\Delta_0}{1 + \alpha(1 - e^{-j\omega_0}z)}\right) \\
&\quad + z\left(\frac{e^{-j\omega_0} - (1+1/\alpha)e^{j\omega_0}}{e^{-j2\omega_0} - 1} \frac{z - e^{j\omega_0}}{z - (1+1/\alpha)e^{j\omega_0}} \frac{(\Delta_0^T)^*}{1 + \alpha(1 - e^{j\omega_0}z)}\right),
\end{aligned}$$

where $\alpha > 0$; here the superscript ^{T} denotes the real adjoint of the operator. Clearly this function satisfies the interpolation condition, and is real since it equals its parahermitian conjugate.

Because $\|\Delta\|_{\mathcal{K}_2 \to \mathcal{K}_2} < 1$, we can choose α sufficiently large so that $\|\check{\Delta}\|_\infty < 1$, satisfying the norm constraint.

To complete the proof we show that $\check{\Delta}$ corresponds to a causal operator $\boldsymbol{\Delta}$ on the space $\mathcal{L}_2$: start by factoring the function

$$\check{\Delta}(z) =: z\check{\Delta}_1(z). \tag{6.2}$$

Now, $\check{\Delta}_1$ corresponds to a causal operator $\tilde{\Delta}_1$ on $\boldsymbol{\ell}_2$ since $\check{\Delta}_1$ is in $\boldsymbol{\mathcal{A}}$. That is, for any $\tilde{w} \in \boldsymbol{\ell}_2$ we have that $(\tilde{\Delta}_1\tilde{w})[k_0]$, for $k_0 \geq 0$, only depends on the values of $\tilde{w}[k]$ for $0 \le k \le k_0$. Therefore, with $\boldsymbol{\Delta}_1 = W^{-1}\tilde{\Delta}_1 W$ we have that, for any $w \in \mathcal{L}_2$, the value of

$$(\boldsymbol{\Delta}_1 w)(t_0) \qquad \text{for } t_0 \ge 0,$$

depends only on the function values of $w(t)$ in the interval $0 \le t \le k_0 h$, where k_0 is the smallest integer so that $t_0 \le k_0 h$. Directly from (6.2) we get that

$$\boldsymbol{\Delta} w = \mathbf{D}_h \boldsymbol{\Delta}_1 w,$$

where $\mathbf{D}_h$ is the delay operator, and so

$$(\boldsymbol{\Delta} w)(t_0) = \begin{cases} 0, & 0 \le t_0 < h \\ (\boldsymbol{\Delta}_1 w)(t_0 - h), & h \le t_0. \end{cases}$$

Therefore, $(\boldsymbol{\Delta} w)(t_0)$ depends only on the values of $w(t)$ on the interval $0 \le t \le (k_0 - 1)h$ where k_0 is the smallest integer satisfying $t_0 \le k_0 h$. That is, $\boldsymbol{\Delta}$ is a causal map. ■

We have avoided the zero and π-valued frequency points in the above result in order to construct a real perturbation. If this restriction is removed we have the following.

Corollary 6.2 *Given* $\Delta \in \mathcal{U}\boldsymbol{\Delta}_{PTV}$ *and* $\omega_0 \in (-\pi, \pi]$. *There exists a causal operator* $\boldsymbol{\Delta} \in \mathcal{U}\mathfrak{X}_{PTV}$, *so that*

$$\check{\Delta}(e^{j\omega_0}) = \Delta.$$

Proof Set $\check{\Delta}(z) = ze^{-j\omega_0}\Delta$; causality follows by the same argument as above. ■

We next decompose the nominal system $\mathbf{M}$ with respect to uncertainty and performance: Let w and z be in $\mathcal{L}_2^r$, and $q_1, q_2 \in \mathcal{L}_2^m$, and therefore $\mathbf{M}$ can be viewed as a mapping from $\mathcal{L}_2^m \oplus \mathcal{L}_2^r$ to $\mathcal{L}_2^m \oplus \mathcal{L}_2^r$ with the partition

$$\mathbf{M} =: \begin{bmatrix} \mathbf{M}^{11} & \mathbf{M}^{12} \\ \mathbf{M}^{21} & \mathbf{M}^{22} \end{bmatrix}.$$

We can now prove the following stabilization result, which refers to the definition of robust stability stated in Definition 3.2.

Theorem 6.3 *Suppose* $\mathbf{M} \in \mathfrak{L}_{\mathcal{A}}$*, and that its transfer function* $\check{M}(z)$ *is a compact operator at every* $z \in \bar{\mathbb{D}}$*. The system in Figure 6.1 is robustly stabilized to* $\mathcal{U}\mathfrak{X}_{PTV}$ *if and only if*

$$\sup_{\omega \in (-\pi, \pi]} \mu_{\mathbf{\Delta}_{PTV}}(\check{M}^{11}(e^{j\omega})) \leq 1. \tag{6.3}$$

Proof By Lemma 3.4 it is sufficient to show that $(\mathbf{I} - \mathbf{M}^{11}\Delta)^{-1}$ exists in $\mathfrak{L}(\mathcal{L}_2)$ for all $\mathbf{\Delta} \in \mathcal{U}\mathfrak{X}_{PTV}$ if and only if (6.3) holds.

(If): By contrapositive: suppose that there exists $\mathbf{\Delta} \in \mathcal{U}\mathfrak{X}_{PTV}$ such that

$$\mathbf{I} - \mathbf{M}^{11}\Delta \quad \text{is singular.}$$

Then by Lemma 4.2 and Lemma 4.4 we have that there exists $\omega_0 \in (-\pi, \pi]$ so that

$$\text{rad}(\check{M}^{11}(e^{j\omega_0})\check{\Delta}(e^{j\omega_0})) \geq 1$$

holds. Now from the definition of the set $\mathbf{\Delta}_{PTV}$ in (6.1) we have that $\check{\Delta}(e^{j\omega_0})$ is in its open unit ball; therefore $\mu_{\mathbf{\Delta}_{PTV}}(\check{M}^{11}(e^{j\omega_0})) > 1$.

(Only if): We again employ the contrapositive. Suppose that (6.3) is not satisfied. Then there exist $\omega_0 \in (-\pi, \pi]$ and $\Delta' \in \mathcal{U}\mathbf{\Delta}_{PTV}$ so that $\text{rad}(\check{M}^{11}(e^{j\omega_0})\Delta') > 1$. The function $\text{rad}(\check{M}(e^{j\omega})\Delta')$ is continuous at ω_0 by Proposition 2.2, since $\check{M}(e^{j\omega_0})\Delta'$ is a compact operator and $\check{M}(e^{j\omega})\Delta'$ is by definition a continuous function. Hence, there is a neighborhood of ω_0 in which $\text{rad}(\check{M}(e^{j\omega})\Delta') > 1$. We therefore assume, without loss of generality, that $\omega_0 \in (-\pi, \pi)$ is nonzero and $\text{rad}(\check{M}^{11}(e^{j\omega_0})\Delta') > 1$.

Now, by scaling Δ' we can obtain a $\Delta \in \mathcal{U}\mathbf{\Delta}_{PTV}$ so that $I - \check{M}^{11}(e^{j\omega_0})\Delta$ is singular. By Lemma 6.1 there exists a causal operator $\mathbf{\Delta} \in \mathcal{U}\mathfrak{X}_{PTV}$, so that

$$\check{\Delta}(e^{j\omega_0}) = \Delta.$$

Hence, we conclude that $I - \check{M}^{11}\check{\Delta}$ is not invertible in $\mathcal{A}$, and therefore by Lemma 4.2 that $\mathbf{I} - \mathbf{M}^{11}\Delta$ is singular. ■

We proceed to convert our robust performance criterion into a pure robust stability criterion. To facilitate this, define the uncertainty set,

$$\boldsymbol{\Delta}_{rp} := \{\Delta = \mathrm{diag}(\Delta_0, \Delta_1) \in \mathfrak{L}(\mathcal{K}_2^{m+r}) : \Delta_0 \in \boldsymbol{\Delta}_{PTV}, \Delta_1 \in \mathfrak{L}(\mathcal{K}_2^r)\}. \quad (6.4)$$

We can now prove the next theorem, which states that robust performance is equivalent to a stabilization condition. The theorem's statement uses the form of robust performance given in Definition 3.3

Theorem 6.4 *Suppose* $\mathbf{M} \in \mathfrak{L}_{\mathfrak{A}}$, *and that its transfer function* $\check{M}(z)$ *is a compact operator at every* $z \in \bar{\mathbb{D}}$. *The system in Figure 6.1 has robust performance to* $\mathcal{U}\mathfrak{X}_{PTV}$ *if and only if*

$$\sup_{\omega \in (-\pi, \pi]} \mu_{\boldsymbol{\Delta}_{rp}}(\check{M}(e^{j\omega})) \leq 1. \quad (6.5)$$

Proof Invoking Theorem 6.3 and Definition 3.3, robust performance to the perturbation set $\mathcal{U}\mathfrak{X}_{PTV}$ is equivalent to

$$\sup_{\omega \in (-\pi, \pi]} \mu_{\boldsymbol{\Delta}_{PTV}}(\check{M}^{11}(e^{j\omega})) \leq 1, \quad (6.6)$$

and that for all $\boldsymbol{\Delta} \in \mathcal{U}\mathfrak{X}_{PTV}$ the equality

$$\|\mathcal{F}_u(\mathbf{M}, \boldsymbol{\Delta})\|_{\mathcal{L}_2 \to \mathcal{L}_2} \leq 1,$$

is satisfied. Because $\mathbf{M}$ is in $\mathfrak{L}_{\mathfrak{A}}$ and $\mathfrak{X}_{PTV}$ is a subspace of $\mathfrak{L}_{\mathfrak{A}}$, the latter condition can be expressed in the frequency domain by the statement that for all $\boldsymbol{\Delta} \in \mathcal{U}\mathfrak{X}_{PTV}$

$$\|\mathcal{F}_u(\check{M}, \check{\Delta})\|_{\infty} \leq 1, \quad (6.7)$$

where $\check{\Delta}$ is the transfer function of $\boldsymbol{\Delta}$. By essentially the same argument given in the "only if" part of the proof of Theorem 6.3, we can show that (6.7) does not hold, if and only if there exist a frequency $\omega_0 \in (-\pi, \pi]$ and a perturbation $\Delta \in \mathcal{U}\boldsymbol{\Delta}_{PTV}$ so that $\|\mathcal{F}_u(\check{M}(e^{j\omega_0}), \Delta)\|_{\mathcal{K}_2 \to \mathcal{K}_2} > 1$. Therefore, the condition in (6.7) is equivalent to

$$\sup_{\omega \in (-\pi, \pi]} \sup_{\Delta \in \mathcal{U}\boldsymbol{\Delta}_{PTV}} \|\mathcal{F}_u(\check{M}(e^{j\omega}), \Delta)\|_{\mathcal{K}_2 \to \mathcal{K}_2} \leq 1. \quad (6.8)$$

We have shown that robust performance is equivalent to the conditions in (6.6) and (6.8). To get the claim, it is routine to use the main loop argument of Lemma 5.14 and show that (6.6) and (6.8) hold if and only if (6.5) holds. ■

In a similar way to the previous chapter we have reduced our robustness specification to a structured singular value condition on the unit circle; here however the uncertainty structure is much simpler since it has a finite number of blocks. The above result is similar to the purely discrete time robust performance problem. The key difference is the infinite dimensionality of the operator $\check{M}(e^{j\omega})$. Next we consider perturbations that are nearly h-periodic.

6.1.2 Quasi-Periodic Perturbations

We now develop robust performance conditions for a larger class of uncertainty, namely operators that are quasi-periodic. Quasi-periodic operators are operators which are close to being h-periodic. The exact performance test which we develop for this uncertainty class has the advantage of not being structured singular condition, and so, as we will see in Section 6.2, can be evaluated from a quasi-convex optimization on Euclidean space. The results and proofs closely follow work by Poolla and Tikku[50] on purely discrete time systems.

Define the class of causal quasi-periodic operators on $\mathcal{L}_2$ by

$$\mathfrak{L}_{QP}(\nu) := \{\boldsymbol{\Delta} \in \mathfrak{L}(\mathcal{L}_2) : \|\mathbf{D}_h\boldsymbol{\Delta} - \boldsymbol{\Delta}\mathbf{D}_h\| < \nu,\ \boldsymbol{\Delta}\text{ causal}\},$$

where $\nu > 0$ and $\mathbf{D}_h$ is the h-delay on $\mathcal{L}_2$. For all ν this family of sets contains the h-periodic operators, and tends to them pointwise as ν tends to zero. For this latter reason we call them quasi-periodic. Define the structured perturbation set

$$\mathfrak{X}_{QP}(\nu) := \mathfrak{L}_{QP}(\nu) \cap \mathfrak{X}_s,$$

where $\mathfrak{X}_s$ is the spatially structured set defined in (3.2).

Before stating the main results of the subsection we require the following set of nonsingular scaling operators:

$$\tilde{\mathcal{D}}_{\Delta_{rp}} := \{D \in \mathfrak{L}\,(\mathcal{K}_2^{m+r}) : D = \mathrm{diag}(d_1 I, \ldots, d_{d+1} I),\ d_k \in \mathbb{C} \text{ and } d_k \neq 0\}, \tag{6.9}$$

where the spatial structure of the set is such that each $D \in \tilde{\mathcal{D}}_{\Delta_{rp}}$ commutes with every member of the uncertainty set $\boldsymbol{\Delta}_{rp}$ defined in (6.4). Therefore, for each real number ω_0, the product $D\check{M}(e^{j\omega_0})D^{-1}$ is well-defined. Note that although the set $\tilde{\mathcal{D}}_{\Delta_{rp}}$ is a subset of $\mathfrak{L}\,(\mathcal{K}_2^m)$, it is clearly isomorphic to a set in Euclidean space.

We can now state two complementary results pertaining to the structured perturbation class $\mathfrak{X}_{QP}(\nu)$. Note that the infimum in the theorems is the natural upper bound, based on the matrix case described in Section 2.6, for the structured singular value $\mu_{\Delta_{rp}}(\check{M}(e^{j\omega}))$ of Subsection 6.1.1.

Theorem 6.5 *Suppose* $\mathbf{M} \in \mathfrak{L}_{\mathcal{A}}$. *If*

$$\sup_{\omega\in(-\pi,\pi]}\ \inf_{D\in\tilde{\mathcal{D}}_{\Delta_{rp}}} \|D\check{M}(e^{j\omega})D^{-1}\| > 1,$$

then the system in Figure 6.1 does not have robust performance to the perturbation sets $\mathcal{U}\mathfrak{X}_{QP}(\nu)$ *for any* $\nu > 0$.

This is in contrast with the periodic case because the theorem does not hold in general if we replace $\mathfrak{X}_{QP}(\nu)$ with $\mathfrak{X}_{PTV}$. We also have the following sufficiency result.

Theorem 6.6 *Suppose* $\mathbf{M} \in \mathfrak{L}_{\mathcal{A}}$. *If*

$$\sup_{\omega\in(-\pi,\pi]}\ \inf_{D\in\tilde{\mathcal{D}}_{\Delta_{rp}}} \|D\check{M}(e^{j\omega})D^{-1}\| < 1,$$

then there exists $\nu > 0$ *so that the system in Figure 6.1 has robust performance to the perturbation set* $\mathcal{U}\mathfrak{X}_{QP}(\nu)$.

These theorems do not cover the case when the minimization is equal to one; we have an immediate corollary that covers this case. For $\nu > 0$ define

$$\rho_\nu := \inf\{\|\boldsymbol{\Delta}\| : \boldsymbol{\Delta} \in \mathfrak{X}_{QP}(\nu),\ 1 \in \operatorname{spec}(\mathbf{M}^{11}\boldsymbol{\Delta}) \text{ or } \|\boldsymbol{\Delta}\| \cdot \|\mathcal{F}_l(\mathbf{M}, \boldsymbol{\Delta})\| > 1\}.$$

In other words ρ_ν is the radius of the largest open ball in $\mathfrak{L}_{QP}(\nu)$ to which the system has robust performance. With this definition we have the following corollary.

Corollary 6.7

$$\lim_{\nu\to 0} \rho_\nu = \sup_{\omega\in(-\pi,\pi]} \inf_{D\in\mathcal{D}_{\Delta rp}} \|D^{\frac{1}{2}}\check{M}(e^{j\omega})D^{-\frac{1}{2}}\|.$$

Of these results we prove Theorem 6.5 which is based on [50]. First, some intermediate results are required. A key result on which the proof is constructed is now given. The result was first proved for matrices in [41]; also see [10]. In the statement of the result, the operator Q on $\mathcal{K}_2$ defines a memoryless multiplication operator on ℓ_2.

Proposition 6.8 *Suppose that $Q \in \mathfrak{L}(\mathcal{K}_2^m)$. If $\inf_{D\in\mathcal{D}_{\Delta rp}} \|D^{\frac{1}{2}}QD^{-\frac{1}{2}}\| > 1$, then there exist nonzero $\tilde{w} \in \ell_2^m$ and $\gamma > 1$, so that with $\tilde{z}[k] = Q\tilde{w}[k]$ the inequalities*

$$\|\tilde{z}_l\|_{\ell_2} \geq \gamma\|\tilde{w}_l\|_{\ell_2} \quad \text{hold for } 1 \leq l \leq d+1,$$

where $(\tilde{w}_1, \ldots, \tilde{w}_{d+1}) = \tilde{w}$, $(\tilde{z}_1, \ldots, \tilde{z}_{d+1}) = \tilde{z}$ and $\tilde{w}_l, \tilde{z}_l \in \ell_2^{m_l}$.

Reference [41] gives a general version of this result for purely continuous time systems with $\mathcal{L}_2$ inputs. A proof for ℓ_2 is given in Appendix G, where we show a more general result than the one stated above, that the so-called S-procedure is lossless for shift invariant quadratic forms. We remark that Bercovici et al. [10] prove a stronger result for the memoryless matrix version of the above claim; similarly the above result can be strengthened so that the sequence $\tilde{w}$ can be chosen to have finite support, with $\gamma = 1$, providing Q is further restricted to be a compact operator.

Next, define the following set of scalar functions with respect to the parameters q and ω_0:

$$\phi^q_{\omega_0}[k] := \begin{cases} e^{j\omega_0 k}, & 0 \le k \le q \\ 0, & q < k \end{cases}$$

Then given any function $a \in \mathcal{K}_2$ we have that

$$a\phi^q_{\omega_0} \in \boldsymbol{\ell}_2,$$

for every $q \ge 0$ and $\omega_0 \in (-\pi, \pi]$. This set of functions has the following well-known and useful property.

Lemma 6.9 *Suppose* $\mathbf{M} \in \mathfrak{L}_{\mathcal{A}}$, $\omega_0 \in (-\pi, \pi]$, *and the function* $a \in \mathcal{K}_2$ *is nonzero. Then with* $\tilde{w}^q := a\phi^q_{\omega_0}$ *the limit*

$$\lim_{q\to\infty} \frac{\|\tilde{M}\tilde{w}^q - \check{M}(e^{j\omega_0})\tilde{w}^q\|_{\ell_2}}{\|\tilde{w}^q\|_{\ell_2}} = 0,$$

where $\check{M}$ *is the transfer function for* $\mathbf{M}$.

The proof is a routine application of the z-transform and the frequency domain. It states a fundamental property of LTI systems, namely that they have the harmonics as their approximate eigenfunctions.

We now have the main lemma used in the proof of Theorem 6.5 to construct a destabilizing perturbation. The result provides a bound on the amount of variation required for an operator to map a set of harmonics *onto* itself maintaining a power inequality.

Lemma 6.10 *Suppose*

(i) $\{\omega_k\}_{k=1}^n$ *is a sequence of* n *distinct frequencies that satisfy* $-\pi < \omega_1 < \ldots < \omega_n < \pi + \omega_1$ *and* $\omega_l \ne -\omega_k$ *for all* k *and* l *in* $\{1, \ldots, n\}$.

(ii) $\{a^k\}_{k=1}^n$ *and* $\{b^k\}_{k=1}^n$ *are sequences in* $\mathcal{K}_2$, *and* $\gamma > 1$, *so that*

$$\gamma \sum_{k=1}^{n} \|b^k\|^2_{\mathcal{K}_2} \le \sum_{l=1}^{n} \|a^k\|^2_{\mathcal{K}_2}.$$

Then there exists a causal $\Delta \in \mathcal{UL}_{QP}(\nu)$*, where* $\nu := 2\sin(\frac{\omega_n - \omega_1}{2})$*, so that*

$$\frac{\|\tilde{\Delta}\tilde{z}^q - \tilde{w}^q\|_{\ell_2}}{\|\tilde{z}^q\|_{\ell_2}} \xrightarrow{q\to\infty} 0,$$

where $\tilde{z}^q := \sum_{k=1}^n a^k \phi_{\omega_k}^q$*,* $\tilde{w}^q := \sum_{k=1}^n b^k \phi_{\omega_k}^q$*.*

The proof is tedious and is therefore relegated to Appendix E. Figure 4.3 can be used to interpret the difference between periodic and quasi-periodic perturbations: given the sequence of harmonic functions $\phi_{\omega_1}, \ldots, \phi_{\omega_n}$ in ℓ_2, a periodic perturbation must map them in a diagonal fashion as shown in the figure. However, the quasi-periodic perturbation in the lemma allows the off diagonal blocks of the picture to be nonzero and thus permits cross coupling between the harmonics.

We can now prove the theorem. The proof is divided into two steps.

Proof of Theorem 6.5

Choose any $\nu > 0$.

Step 1: construct a candidate destabilizing perturbation Δ

By hypothesis there exists a frequency $\omega^0 \in (-\pi, \pi]$ so that

$$\inf_{D \in \mathcal{D}_{\Delta rp}} \|D\check{M}(e^{j\omega^0})D^{-1}\| > 1.$$

Therefore, by Proposition 6.8 there exist $\gamma > 1$ and $\tilde{b} = (\tilde{b}_1, \ldots, \tilde{b}_{d+1}) \in \ell_2^m$, with each $\tilde{b}_k \in \ell_2^{m_k}$, so that with $\tilde{a}' = (\tilde{a}'_1, \ldots, \tilde{a}'_{d+1}) := \check{M}(e^{j\omega^0})\tilde{b}$ we have

$$\gamma\|\tilde{b}_l\|_{\ell_2} \le \|\tilde{a}'_l\|_{\ell_2} = \sqrt{\sum_{k=0}^{\infty} \|(\check{M}(e^{j\omega^0})\tilde{b}[k])_l\|_{\mathcal{K}_2}^2} \tag{6.10}$$

for $1 \le l \le d+1$.

Because $\check{M}(e^{j\omega_0})$ is a memoryless operator on ℓ_2 we may assume, without loss of generality, that $\tilde{b}$ has finite support and (6.10) holds; let n be the support length of $\tilde{b}$.

Now, from (6.10) and the continuity of $\check{M}$ we can choose n distinct frequencies $-\pi < \omega_1 < \ldots < \omega_n < \pi$ in a neighborhood of ω_0, and $0 < \gamma' < \gamma$, so that defining

$$\tilde{a}[k] := \check{M}(e^{j\omega_k})\tilde{b}[k], \tag{6.11}$$

we have the inequalities

$$\gamma' \sqrt{\sum_{k=0}^{n} \|\tilde{b}_l[k]\|_{\mathcal{K}_2}^2} = \gamma' \|\tilde{b}_l\|_{\ell_2} \le \|\tilde{a}_l\|_{\ell_2} = \sqrt{\sum_{k=0}^{n} \|(\check{M}(e^{j\omega_k})\tilde{b}[k])_l\|_{\mathcal{K}_2}^2} \quad (6.12)$$

satisfied for each l in $\{1, \dots, d+1\}$. Furthermore, these frequencies can be chosen so that $\nu > 2\sin(\frac{\omega_n - \omega_1}{2})$ and hypothesis (i) of Lemma 6.10 holds.

Also, note that by (6.12), for each fixed $l \in \{1, \dots, d+1\}$, the sequences $\tilde{a}_l[k]$ and $\tilde{b}_l[k]$ are of length n and satisfy (ii) in Lemma 6.10. We can therefore apply the lemma.

Define the sequence of functions in ℓ_2 for $q \ge 0$ by

$$\tilde{z}_l^q := \sum_{k=1}^{n} \tilde{a}_l[k]\phi_{\omega_k}^q, \qquad \tilde{w}_l^q := \sum_{k=1}^{n} \tilde{b}_l[k]\phi_{\omega_k}^q, \quad (6.13)$$

for each $l \in \{1, \dots, d+1\}$.

Thus by Lemma 6.10 there exists an operator $\Delta_l \in \mathcal{UL}_{QP}(\nu)$, for each $l \in \{1, \dots, d+1\}$, so that

$$\frac{\|\tilde{\Delta}_l \tilde{z}_l^q - \tilde{w}_l^q\|_{\ell_2}}{\|\tilde{z}_l^q\|_{\ell_2}} \xrightarrow{q \to \infty} 0. \quad (6.14)$$

Our candidate perturbation is $\Delta = \operatorname{diag}(\Delta_1, \dots, \Delta_{d+1})$ which is now in $\mathcal{UX}_{QP}(\nu)$ by construction.

Step 2: Demonstrate that Δ is destabilizing

Using the definitions in (6.13) we first set $\tilde{w}^q = (\tilde{w}_1^q, \dots, \tilde{w}_{d+1}^q)$ and $\tilde{z}^q = (\tilde{z}_1^q, \dots, \tilde{z}_{d+1}^q)$. From (6.14) we therefore have

$$\frac{\|\tilde{\Delta}\tilde{z}^q - \tilde{w}^q\|_{\ell_2}}{\|\tilde{z}^q\|_{\ell_2}} \xrightarrow{q \to \infty} 0. \quad (6.15)$$

Since $\|\tilde{M}\| \cdot \|\tilde{\Delta}\tilde{z}^q - \tilde{w}^q\|_{\ell_2} \ge \|\tilde{M}\tilde{\Delta}\tilde{z}^q - \tilde{M}\tilde{w}^q\|_{\ell_2}$ we get the limit

$$\frac{\|\tilde{M}\tilde{\Delta}\tilde{z}^q - \tilde{M}\tilde{w}^q\|_{\ell_2}}{\|\tilde{z}^q\|_{\ell_2}} \xrightarrow{q \to \infty} 0. \quad (6.16)$$

Now, using the definition of $\tilde{z}_l^q$ given in (6.11) and (6.13), and Lemma 6.9 it follows that

$$\frac{\|\tilde{M}\tilde{w}^q - \tilde{z}^q\|_{\ell_2}}{\|\tilde{w}^q\|_{\ell_2}} \xrightarrow{q \to \infty} 0.$$

It is also routine to verify from (6.12) that

$$\lim_{q\to\infty} \frac{\|\tilde{w}^q\|_{\ell_2}}{\|\tilde{z}^q\|_{\ell_2}} < \infty.$$

So,

$$\frac{\|\tilde{M}\tilde{w}^q - \tilde{z}^q\|_{\ell_2}}{\|\tilde{z}^q\|_{\ell_2}} \overset{q\to\infty}{\longrightarrow} 0.$$

With the last limit and (6.16) we invoke the triangle inequality to get

$$\frac{\|\tilde{M}\tilde{\Delta}\tilde{z}^q - \tilde{z}^q\|_{\ell_2}}{\|\tilde{z}^q\|_{\ell_2}} \overset{q\to\infty}{\longrightarrow} 0.$$

This means that $1 \in \text{spec}(\tilde{M}\tilde{\Delta})$ since it is an approximate eigenvalue. Hence, $\mathbf{I} - \mathbf{M\Delta}$ is not invertible in $\mathfrak{L}(\mathcal{L}_2)$. Since $\mathbf{\Delta} \in \mathcal{U}\mathfrak{X}_{QP}(\nu)$ and ν was arbitrary the claim follows. ∎

We omit the proof of Theorem 6.6, which is based on the the small gain condition, as it is a routine translation of the discrete time case shown in [50] to our operator theoretic framework. The basic idea behind the proof is to first show that $\sup_{\omega\in(-\pi,\pi]} \inf_{D\in\tilde{\mathcal{D}}_{\Delta rp}} \|D\check{M}(e^{j\omega})D^{-1}\| = \inf_{\check{D}\in\check{\mathcal{D}}_{\Delta rp}} \|\check{D}\check{M}\check{D}^{-1}\|_\infty$, where $\check{\mathcal{D}}_{\Delta rp}$ is the set of invertible functions in $\mathcal{A}$ mapping $\bar{\mathbb{D}}$ to $\tilde{\mathcal{D}}_{\Delta_{rp}}$. Every $\check{D}$ and $\check{D}^{-1}$ (ignoring phase) can be approximated as closely as desired by polynomials, and therefore for $\nu > 0$ sufficiently small, these approximations and a small gain type argument similar to the proof in Proposition 2.15 can be employed. We remark that the validity of Theorem 6.6 is critically dependent on $\check{M}$ being in $\mathcal{A}$; if the transfer function of the nominal system is permitted to have discontinuities on $\partial\mathbb{D}$, then the result does *not* remain valid in general.

In this section we have provided a nonconservative condition for robust performance to quasi-periodic perturbations. Section 6.2 is devoted to computational aspects of evaluating this condition; we show that it can be formulated as a quasi-convex program. We now discuss robust performance to arbitrary time-varying perturbations.

6.1.3 Arbitrary Time-Varying Uncertainty

Two equivalent conditions will be presented which are necessary and sufficient for robust performance of the system in Figure 6.1 to the causal perturbations in $\mathcal{UX}_s$, namely the class of causal, structured and arbitrary time-varying uncertainty. The results are analogous to the continuous time results in [58][41], and follow from minor modifications to the purely discrete-time proofs in [50] (which only rely on the earlier work of [10]) combined with the results of the last subsection on quasi-periodic uncertainty. For this reason we do not prove the results, but merely describe the central ideas behind the proofs. Note however, if we remove the requirement of causality below, the result follows directly from Corollary G.5 of Appendix G, which is proved.

Theorem 6.11 *Suppose* $\mathbf{M} \in \mathfrak{L}_{\mathcal{A}}$. *The system in Figure 6.1 has robust performance to the causal perturbations in* $\mathcal{UX}_s$ *if and only if*

$$\inf_{D \in \tilde{\mathcal{D}}_{\Delta rp}} \sup_{\omega \in (-\pi, \pi]} \|D\check{M}(e^{j\omega})D^{-1}\| \le 1.$$

That the inequality in the theorem guarantees robust performance follows by a standard small gain argument by first noting that every $D \in \tilde{\mathcal{D}}_{\Delta_{rp}}$ commutes with all elements of $\mathcal{UX}_s$. To show the above condition is also necessary for robust performance is more complicated. The first step in the proof (see [50]) is to show that if the condition is violated there exist frequencies $\omega^0, \ldots, \omega^{d+1}$ in $(-\pi, \pi]$, so that there is no $D \in \tilde{\mathcal{D}}_{\Delta_{rp}}$ which satisfies the $d+2$ inequalities $\|D\check{M}(e^{j\omega^k})D^{-1}\| \le 1$. Having established the existence of these frequencies, minor modifications of the proof of Theorem 6.5 yield the result: using a lifted version of Proposition 6.8, we can find, for some $n > 0$, frequencies $\omega_0, \ldots, \omega_n \in (-\pi, 0) \cup (0, \pi)$ and finite sequences $\tilde{a}[k]$ and $\tilde{b}[k]$ in $\mathcal{K}_2$ satisfying (6.11) and (6.12). Here, however these frequencies are in neighborhoods of $\omega^0, \ldots, \omega^{d+1}$, not just in the neighborhood of a single frequency ω^0 as in the proof Theorem 6.5. With the relations (6.11) and (6.12) now in place, the remainder of the proof follows that of Theorem 6.5 exactly, where the resulting destabilizing perturbation is in $\mathcal{UX}_{QP}(\nu)$ with ν set to $\max_{k,l} 2\sin(\frac{\omega_k-\omega_l}{2})$.

The scaling condition in the last theorem requires that the minimization be conducted taking all frequencies in the interval $(-\pi,\pi]$ into account. If we assume that the transfer function $\check{M}(z)$ has a form such as that arising from Figure 3.1, then the test can be reformulated in a simple way. Note that in the following theorem statement $\|\cdot\|$ refers to the $\mathbb{C}^{\tilde{n}} \oplus \mathcal{K}_2 \to \mathbb{C}^{\tilde{n}} \oplus \mathcal{K}_2$ induced norm.

Theorem 6.12 *Suppose* $\mathbf{M} \in \mathfrak{L}_{\mathcal{A}}$ *and that* $\check{M}(z) := \check{C}z(I - A_d z)^{-1}\check{B} + \check{D}$ *as in (3.5). Then the system in Figure 6.1 has robust performance to the causal perturbations in* $\mathcal{UX}_s$ *if and only if*

$$\inf_{D_0\in\tilde{\mathcal{D}}_0,\, D\in\tilde{\mathcal{D}}_{\Delta rp}} \left\| \begin{bmatrix} D_0 & 0 \\ 0 & D \end{bmatrix} \begin{bmatrix} A_d & \check{B} \\ \check{C} & \check{D} \end{bmatrix} \begin{bmatrix} D_0^{-1} & 0 \\ 0 & D^{-1} \end{bmatrix} \right\| \le 1, \qquad (6.17)$$

where the set $\tilde{\mathcal{D}}_0 := \{D_0 \in \mathbb{C}^{\tilde{n}\times\tilde{n}} : D_0 \text{ is nonsingular}\}$.

This result states that $\inf_{D\in\tilde{\mathcal{D}}_{\Delta rp}} \sup_{\omega\in(-\pi,\pi]} \|D\check{M}(e^{j\omega})D^{-1}\| \le 1$ is equivalent to inequality (6.17), and is analogous to a discrete-time result. Since the proof of the above theorem would take us quite far afield, and involves a routine generalization of the matrix proof in [49], we simply describe the main steps in the proof, and refer the reader to [49] for more details. First, a main loop argument similar to Lemma 5.14 is used to show that inequality (6.17) holds if and only if the structured singular value inequality

$$\inf_{D\in\tilde{\mathcal{D}}_{\Delta rp}} \mu_{\boldsymbol{\Delta}}\left(\begin{bmatrix} A_d & \check{B}D^{-1} \\ D\check{C} & D\check{D}D^{-1} \end{bmatrix} \right) \le 1 \quad \text{is satisfied,}$$

where $\boldsymbol{\Delta} := \{\mathrm{diag}(zI, \Delta) : z \in \mathbb{C} \text{ and } \Delta \in \mathfrak{L}(\mathcal{K}_2)\}$. Now, using a generalized version of a proof in [49], it is possible to show that the block structure $\boldsymbol{\Delta}$ is so-called μ-simple; this means that

$$\mu_{\boldsymbol{\Delta}}\left(\begin{bmatrix} A_d & \check{B}D^{-1} \\ D\check{C} & D\check{D}D^{-1} \end{bmatrix} \right) = \inf_{D_0\in\tilde{\mathcal{D}}_0} \left\| \begin{bmatrix} D_0A_dD_0^{-1} & D_0\check{B}D^{-1} \\ D\check{C}D_0^{-1} & D\check{D}D^{-1} \end{bmatrix} \right\|$$

for each $D \in \tilde{\mathcal{D}}_{\Delta rp}$. Hence (6.17) follows immediately.

We have now come to the end of this section, where necessary and sufficient conditions for robust performance to three different types of structured uncertainty have been derived. The obtained conditions look analogous to well-known results on the corresponding discrete time problems. If *unstructured* robust stability is considered for each of these time-varying uncertainty classes, it is straightforward to verify that all three conditions reduce to the same small gain condition; this contrasts significantly with the LTI robust stabilization results of Chapter 4.

Next we show that evaluating the performance conditions in Theorems 6.5 and 6.6, and Theorem 6.17 can be accomplished by quasi-convex optimization on Euclidean space; this fact can also be seen in more explicit form in (6.30) of Subsection 6.3.1. The optimization problems, as the conditions developed here, are in terms of the operator $\check{M}(e^{j\omega_0})$; therefore tools for computing with this operator are also developed.

6.2 Computational Tools

Our goal now is to investigate computational aspects of the robust performance conditions of the preceding section. We will concentrate on the conditions constructed for quasi-periodic uncertainty in Subsection 6.1.2. The corresponding computations for the performance condition to arbitrary time-varying uncertainty presented in Theorem 6.12 follow by routine modifications to the tools developed here.

Computation of the structured singular value condition in 6.1.1 will not be directly addressed, since exact computational schemes are still being sought in the matrix case. We, however, point out that the scaling condition in Section 6.1.2 provides the standard upper bound for this structured singular value; such scaling conditions were first introduced for matrices in [20] and [56], where they are exact for some uncertainty structures. See [49] for a synopsis.

The section is divided as follows: the first subsection develops some of the general computational properties of the infimization and converts it to a more convenient form. The next one concentrates on explicitly checking these general properties for the sampled-data transfer function.

6.2.1 Definition and Properties

The results of this section are concerned with the infimization of Subsection 6.1.2, which in addition to being an exact robust performance condition for quasi-periodic uncertainty, is also a generalization to compact operators of the usual upper bound for the structured singular value of a matrix. We concentrate on the infimization of 6.1.2, however the results here routinely extend to the infimization condition in Theorem 6.12, and closely follow the matrix case; see for example [49] or [72].

Specifically, we are concerned with the computational properties of $\inf_{D\in\tilde{\mathcal{D}}_{\Delta rp}} \|D\check{M}(e^{j\omega_0})D^{-1}\|$ at a fixed frequency ω_0. In the sequel, rather than working with $\check{M}(e^{j\omega_0})$ directly, it is more convenient to consider a compact operator M on an arbitrary Hilbert space. Note that the operator in Theorem 6.12 is also compact. We take this approach and relate our results to the concrete object $\inf_{D\in\tilde{\mathcal{D}}_{\Delta rp}} \|D\check{M}(e^{j\omega_0})D^{-1}\|$ along the way.

Our first step is to establish some new notation. Let $\mathcal{H}$ be a separable Hilbert space and $\boldsymbol{\Delta}$ a subspace of $\mathfrak{L}(\mathcal{H})$. Define the set $\boldsymbol{\Delta}$ of nonsingular commuting operators by

$$\tilde{\mathcal{D}}_{\boldsymbol{\Delta}} := \{D \in \mathfrak{L}(\mathcal{H}) : D \text{ is nonsingular and } D\Delta = \Delta D \text{ for all } \Delta \in \boldsymbol{\Delta}\}.$$

With these definitions in place, we focus throughout on the infimization

$$\inf_{D\in\tilde{\mathcal{D}}_{\boldsymbol{\Delta}}} \|DMD^{-1}\|. \tag{6.18}$$

Note that it is the infimum over a set of weighted norms on M. Clearly it includes $\inf_{D\in\tilde{\mathcal{D}}_{\Delta rp}} \|D\check{M}(e^{j\omega_0})D^{-1}\|$ as a special case.

The next lemma provides a more convenient way of expressing norm conditions in terms of inequalities which are linear in D^*D. It also shows that the set $\tilde{\mathcal{D}}_{\boldsymbol{\Delta}}$ can be made smaller without affecting the above infimum.

Proposition 6.13 *Given an operator M in $\mathfrak{L}(\mathcal{H})$ and $D \in \tilde{\mathcal{D}}_\Delta$. Then for a scalar $\delta > 0$ the following are equivalent.*

(i) $\|DMD^{-1}\|^2 \leq \delta$

(ii) $M^*D^*DM - \delta D^*D \leq 0.$

This follows directly from the definitions of the norm of an operator and an operator being nonpositive.

In the sequel we primarily work with the condition in (ii) above. It is therefore convenient to use a set of scaling matrices so that we get a linear relationship in D rather than D^*D. To accomplish this define

$$\mathcal{D}_\Delta := \{D_0 \in \mathfrak{L}(\mathcal{H}) : D_0 = D^*D \text{ for some } D \in \tilde{\mathcal{D}}_\Delta\}.$$

Now given any nonsingular operator D in $\mathfrak{L}(\mathcal{H})$, there exists a positive and invertible operator D_1 so that

$$D_1^2 = D^*D$$

is satisfied. See for example [71, p. 142]. Hence, the square root operation is well-defined on all elements of $\mathcal{D}_\Delta$. Having defined the set $\mathcal{D}_\Delta$, we have the following corollary which follows directly from the above proposition by setting $D^*D = D_0$.

Corollary 6.14 *Suppose that M is an operator on $\mathcal{H}$. Then*

$$\inf_{D \in \tilde{\mathcal{D}}_\Delta} \|DMD^{-1}\| = \inf_{D \in \mathcal{D}_\Delta} \|D^{\frac{1}{2}} M D^{-\frac{1}{2}}\|.$$

The corollary states that the infimum of the scalings is unaffected if we replace $\tilde{\mathcal{D}}_\Delta$ with the smaller scaling set $\mathcal{D}_\Delta$.

Our new sampled-data scaling set, replacing $\tilde{\mathcal{D}}_{\Delta_{rp}}$, is therefore given by

$$\mathcal{D}_{\Delta_{rp}} := \{D \in \mathfrak{L}(\mathcal{K}_2^{m+r}) : D = \text{diag}(d_1 I, \ldots, d_{d+1} I),\ 0 < d_k \in \mathbb{R}\}.$$

This set of scaling operators is the exact set that is used in the purely continuous time case on a problem with our spatial structure.

So far we have demonstrated that it is sufficient to study $\inf_{D\in\mathcal{D}_\Delta} \|D^{\frac{1}{2}}MD^{-\frac{1}{2}}\|$. We now concentrate on determining, given a $D \in \mathcal{D}_\Delta$, whether the infimum is achieved. The path taken is to examine the linear characterization of the bound, $M^*DM - \delta D$, and attempt to make $\delta > 0$ as small as possible by varying D inside the set $\mathcal{D}_\Delta$.

We require a new object: any self adjoint operator E on $\mathcal{H}$ has a real spectrum, if it also has at least one eigenvalue; we define

$$\lambda_{\max}(E) := \sup\{\lambda \in \operatorname{spec}(E) : \text{there exists } x \in \mathcal{H} \text{ and } Ex = \lambda x\},$$

to be the supremum of the eigenvalues. Next we characterize the largest δ for which our operator inequality holds.

Lemma 6.15 *Suppose that M is a compact operator on $\mathcal{H}$, and $D \in \mathcal{D}_\Delta$. Then*

$$\lambda_{max}(D^{-\frac{1}{2}}M^*DMD^{-\frac{1}{2}}) = \min\{\delta > 0 : M^*DM - \delta D \leq 0\}.$$

Furthermore, there exists an eigenvalue of $D^{-\frac{1}{2}}M^*DMD^{-\frac{1}{2}}$ that achieves the above equality.

Proof By Proposition 6.13 we have that $\min\{\delta > 0 : M^*DM - \delta D \leq 0\} = \|D^{\frac{1}{2}}MD^{-\frac{1}{2}}\|^2$, which is equal to $\|D^{-\frac{1}{2}}M^*DMD^{-\frac{1}{2}}\|$. Now, since M is compact so is $D^{-\frac{1}{2}}M^*DMD^{-\frac{1}{2}}$; it is also self-adjoint and therefore its norm is equal to its spectral radius ([11, p. 199]). That is, $\lambda_{\max}(D^{-\frac{1}{2}}M^*DMD^{-\frac{1}{2}}) = \|D^{-\frac{1}{2}}M^*DMD^{-\frac{1}{2}}\|$, since every nonzero element in the spectrum of a compact operator is an eigenvalue, and the spectrum of $D^{-\frac{1}{2}}M^*DMD^{-\frac{1}{2}}$ contains only nonnegative elements.

That the maximum eigenvalue is achieved follows because the spectrum of a compact operator can only accumulate at 0 [11, p. 193]. ■

Our next step is to examine the behavior of $M^*DM - D$ as D is varied. From now on we assume that $\boldsymbol{\Delta}$ is such that $\mathcal{D}_\Delta$ is a *convex set*. Note that $\mathcal{D}_{\Delta_{rp}}$ is clearly convex.

Define the set of tangent directions at a point $D_0 \in \mathcal{D}_\Delta$ to be the set

$$T_{D_0}(\mathcal{D}_\Delta) := \{Y \in \mathfrak{L}(\mathcal{H}) : \text{there exists } \epsilon_0 > 0 \text{ so that for all } \epsilon \in [0, \epsilon_0),\ D_0 + \epsilon Y \in \mathcal{D}_\Delta\}.$$

That is, all the allowable directions that we can move away from D_0 and remain inside $\mathcal{D}_\Delta$. It is straightforward to verify for each $D_0 \in \mathcal{D}_{\Delta_{rp}}$ that

$$T_{D_0}(\mathcal{D}_{\Delta_{rp}}) := \{Y = \mathrm{diag}(y_1 I, \ldots, y_{d+1} I) \in \mathfrak{L}(\mathcal{K}_2^{m+r}) : y_k \in \mathbb{R}\}.$$

We now develop some results to answer the following question: given a $D_0 \in \mathcal{D}_\Delta$, does there exist a $D \in \mathcal{D}_\Delta$ so that $\|D^{\frac{1}{2}} M D^{-\frac{1}{2}}\| < \|D_0^{\frac{1}{2}} M D_0^{-\frac{1}{2}}\|$? Namely, is D_0 a minimizing weight? The following states that there are no nonglobal minima of quantity $\|D^{\frac{1}{2}} M D^{-\frac{1}{2}}\|$.

Proposition 6.16 *Suppose that M is a compact operator on $\mathcal{H}$ and that D_0 is in the convex set $\mathcal{D}_\Delta$. Then*

$$\inf_{D \in \mathcal{D}_\Delta} \|D^{\frac{1}{2}} M D^{-\frac{1}{2}}\|^2 < \min\{\delta > 0 : M^* D_0 M - \delta D_0 \leq 0\} =: \delta_0$$

if and only if there exists $Y \in T_{D_0}(\mathcal{D}_\Delta)$ so that for $\epsilon > 0$ sufficiently small, the inequality

$$M^*(D_0 + \epsilon Y)M - \delta_0(D_0 + \epsilon Y) < 0 \quad \textit{holds.}$$

Proof (Only if): By Proposition 6.13 if D_0 is suboptimal there necessarily exists a $D_1 \in \mathcal{D}_\Delta$ so that

$$M^* D_1 M - \delta_0 D_1 < 0. \tag{6.19}$$

So for any $\epsilon > 0$ we have that

$$M^*(D_0 + \epsilon D_1)M - \delta_0(D_0 + \epsilon D_1) < 0,$$

which proves this direction because, by convexity of $\mathcal{D}_\Delta$, we have $D_1 \in T_{D_0}(\mathcal{D}_\Delta)$.

(If): For sufficiently small $\epsilon > 0$, the map $D_1 := D_0 + \epsilon Y$ is in $\mathcal{D}_\Delta$ and also satisfies (6.19); hence we can replace δ_0 in (6.19) with a smaller quantity and still satisfy the inequality. The claim then follows by again invoking Proposition 6.13. ■

Hence, we conclude that in order to determine whether we are at an optimal D_0, it is enough to check whether there exists a direction of descent. We now have a key lemma that helps us characterize descent directions.

Lemma 6.17 *Suppose that Q and V are self-adjoint operators on $\mathcal{H}$, and that Q is compact and satisfies $Q \le I$. Then $Q + \epsilon V < I$ for sufficiently small $\epsilon > 0$, if and only if $\langle \psi, V\psi \rangle < 0$ for all nonzero $\psi \in \ker(Q - I)$.*

Proof Because $Q - I$ is self-adjoint, we can partition it with respect to the null space $\ker(Q - I)$ and its orthogonal complement to get

$$\begin{aligned} (Q - I) + \epsilon V &= \begin{bmatrix} Q_+ - I & 0 \\ 0 & 0 \end{bmatrix} + \epsilon \begin{bmatrix} V_1 & V_2 \\ V_2^* & V_3 \end{bmatrix} \\ &= \begin{bmatrix} (Q_+ - I) + \epsilon V_1 & \epsilon V_2 \\ \epsilon V_2^* & \epsilon V_3 \end{bmatrix}. \end{aligned} \tag{6.20}$$

First observe that $Q_+ - I$ is invertible; this is because Q is compact and the spectrum of a compact operator can only cluster at 0. Clearly, $Q_+ - I < 0$.

By continuity of the inverse we may conclude the existence $\epsilon > 0$ sufficiently small so that $(Q_+ - I + \epsilon V_1)^{-1}$ exists and is negative.

We can now prove the result easily using the Schur complement: the operator in (6.20) is negative if and only if

$$\epsilon V_2^* (Q_+ - I + \epsilon V_1)^{-1} V_2 > V_3.$$

See for example [71, p. 153]. The LHS is nonpositive and therefore V_3 must be negative. ■

We are now in a position to prove the following result regarding optimality of a scaling D_0. It states that it is sufficient to focus on a particular subspace of $\mathcal{H}$.

Proposition 6.18 *Suppose M is a compact operator on $\mathcal{H}$, D_0 is in the convex set $\mathcal{D}_\Delta$ and satisfies*

$$M^* D_0 M - D_0 \leq 0$$

and that $\ker(M^* D_0 M - D_0)$ *is* nontrivial. *Then* $\|D_0^{\frac{1}{2}} M D_0^{-\frac{1}{2}}\| > \inf_{D \in \mathcal{D}_\Delta} \|D^{\frac{1}{2}} M D^{-\frac{1}{2}}\|$ *if and only if there exists $Y \in T_{D_0}(\mathcal{D}_\Delta)$ so that*

$$\langle \psi, (M^* Y M - Y)\psi \rangle < 0, \quad \text{for all nonzero } \psi \in \ker(M^* D_0 M - D_0). \quad (6.21)$$

Proof By Proposition 6.16, D_0 is suboptimal if and only if there exists $Y \in T_{D_0}(\mathcal{D}_\Delta)$ so that for sufficiently small $\epsilon > 0$ we have

$$M^*(D_0 + \epsilon Y)M - (D_0 + \epsilon Y) < 0.$$

Clearly this holds if and only if

$$D_0^{-\frac{1}{2}} M^* D_0 M D_0^{-\frac{1}{2}} - I + \epsilon D_0^{-\frac{1}{2}} (M^* Y M - Y) D_0^{-\frac{1}{2}} < 0.$$

We now use Lemma 6.17 to get that the above holds if and only if

$$\langle \phi, D_0^{-\frac{1}{2}} (M^* Y M - Y) D_0^{-\frac{1}{2}} \phi \rangle < 0$$

for all nonzero $\phi \in \ker(D_0^{-\frac{1}{2}} M^* D_0 M D_0^{-\frac{1}{2}} - I)$. This is equivalent to the condition in (6.21). ■

We now have the following lemma which states a well-known fact about the spectral subspaces of compact operators.

Lemma 6.19 *Suppose that M is a compact operator and $\|M\| = 1$. Then there exists a linear mapping $U : \mathbb{C}^u \to \mathcal{H}$, for some integer dimension u, so that*

$$\ker(M^* M - I) = \operatorname{Im} U.$$

Proof Since $M^* M$ is both compact and self-adjoint it is clear that $\lambda_{\max}(M^* M) = 1$ is its spectral radius; see for example [11, p. 199]. Now, the multiplicity of this maximum eigenvalue is finite [11, p. 193], and therefore the corresponding eigenspace is isomorphic to $\mathbb{C}^u$ for some dimension u; hence a mapping U exists as defined in the claim. ■

To complete this section we have a corollary that provides the basis of a constructive finite dimensional optimality test for a $D_0 \in \mathcal{D}_\Delta$. It reduces optimality to a matrix test.

Corollary 6.20 *Suppose that M is a compact operator on $\mathcal{H}$ and that $\|M\| = 1$. Then $\|M\| > \inf_{D\in\mathcal{D}_\Delta} \|D^{\frac{1}{2}} M D^{-\frac{1}{2}}\|$ if and only if there exists $Y \in T_{D_0}(\mathcal{D}_\Delta)$ so that the matrix*

$$U^*(M^*YM - Y)U < 0,$$

where U is a mapping as in Lemma 6.19.

Proof From Lemma 6.19 it is easy to see that the image of U is equal to $\ker(M^*M - I)$; the claim then follows from directly from Proposition 6.18 by setting $D_0 = I$. ■

We have developed the basic optimality properties of the infimization $\inf_{D\in\mathcal{D}_\Delta} \|D^{\frac{1}{2}} M D^{-\frac{1}{2}}\|$. The next section addresses the application of this setup to our sampled-data system.

6.2.2 Reduction to Finite Dimensions

We now move on to developing a way to compute the norm of the scaled transfer function $D^{\frac{1}{2}} \check{M} D^{-\frac{1}{2}}$, and its spectral subspaces. This method can then be applied to compute the objects required in 6.2.1. We also use these calculations in the sequel to develop an algorithm to minimize $\|D^{\frac{1}{2}} \check{M} D^{-\frac{1}{2}}\|$.

Recall from (3.5) that the transfer function has the form

$$\check{M}(z) = \check{C}z(I - zA_d)^{-1}\check{B} + \check{D}. \tag{6.22}$$

We fix z at some $z_0 \in \partial\mathbb{D}$ throughout and omit it in most of our notation.

The results presented in the last section characterized the optimality of a particular scaling operator $D_0 \in \mathcal{D}_{\Delta_{rp}}$. Our present objective is to reduce this optimality test to one that can be *computed* for any $D_0 \in \mathcal{D}_{\Delta_{rp}}$.

Given a D_0 we can easily obtain an expression for $D_0^{\frac{1}{2}} M D_0^{-\frac{1}{2}}$ in terms of the state space data of our nominal system. See Appendix D. A key feature of this scaled operator is that it preserves the structure of (6.22). Therefore, we absorb $D_0^{\frac{1}{2}}$ into $\check{M}$ and then may assume without loss of generality that $D_0 = I$.

Our first goal is to obtain a way to compute $\lambda_{\max}(\check{M}^*\check{M})$. We start by defining the operators $\tilde{P}$ and Q:

$$\begin{aligned}(\check{M}^*\check{M})(z_0) &= [\check{D}^*\check{C} \quad \check{B}^*]\begin{bmatrix} -\check{C}^*\check{C} & Iz_0 - A_d^* \\ Iz_0^* - A_d & 0 \end{bmatrix}^{-1}\begin{bmatrix} \check{C}^*\check{D} \\ \check{B} \end{bmatrix} + \check{D}^*\check{D} \\ &=: \tilde{P}Q\tilde{P}^* + \check{D}^*\check{D}. \end{aligned} \tag{6.23}$$

Also, for a scalar λ define the resolvent

$$\tilde{R}_\lambda := (I\lambda - \check{D}^*\check{D})^{-1}, \tag{6.24}$$

when λ is not in the spectrum of $\check{D}^*\check{D}$. With these definitions, we have the following result that provides a method to compute the maximum eigenvalue of $\check{M}^*\check{M}$ to any desired degree of accuracy.

Proposition 6.21 *Suppose that the scalar $\lambda > \|\check{D}\|^2$, and the matrix $\bar{P}$ satisfies $\bar{P}^*\bar{P} = \tilde{P}^*\tilde{R}_\lambda\tilde{P}$. Then $\lambda_{max}(\check{M}^*\check{M}) < \lambda$ if and only if $0 < I - \bar{P}Q\bar{P}^*$.*

We can use this result to compute $\lambda_{\max}(\check{M}^*\check{M})$ via bisection, with the technical assumption that $\lambda > \|\check{D}\|^2$. See [75] and [28] for details on checking this condition.

Proof Since $\check{M}$ is compact, by Lemma 6.15, we have that $\lambda_{\max}(\check{M}^*\check{M}) < \lambda$ if and only if

$$0 < I\lambda - \check{M}^*\check{M} = I\lambda - \check{D}^*\check{D} - \tilde{P}Q\tilde{P}^*.$$

Now, $\tilde{R}_\lambda > 0$ and hence we can premultiply and postmultiply by its square root to get the equivalent condition

$$0 < I - \tilde{R}_\lambda^{\frac{1}{2}}\tilde{P}Q\tilde{P}^*\tilde{R}_\lambda^{\frac{1}{2}}.$$

This is equivalent to the condition that $\langle \psi, \psi \rangle_{\mathcal{K}_2} > \langle \psi, (\tilde{R}_\lambda^{\frac{1}{2}} \tilde{P} Q \tilde{P}^* \tilde{R}_\lambda^{\frac{1}{2}}) \psi \rangle_{\mathcal{K}_2}$ for all nonzero $\psi \in \mathrm{Im} \tilde{R}_\lambda^{\frac{1}{2}} \tilde{P}$. It is a standard fact that the image space $\mathrm{Im} \tilde{R}_\lambda^{\frac{1}{2}} \tilde{P} = (\ker \tilde{P}^* \tilde{R}_\lambda^{\frac{1}{2}})^{\perp}$. Therefore we can replace the above inequality with

$$\langle \tilde{R}_\lambda^{\frac{1}{2}} \tilde{P} x, (\tilde{R}_\lambda^{\frac{1}{2}} \tilde{P} Q \tilde{P}^* \tilde{R}_\lambda^{\frac{1}{2}}) \tilde{R}_\lambda^{\frac{1}{2}} \tilde{P} x \rangle_{\mathcal{K}_2} < \langle \tilde{R}_\lambda^{\frac{1}{2}} \tilde{P} x, \tilde{R}_\lambda^{\frac{1}{2}} \tilde{P} x \rangle_{\mathcal{K}_2},$$

for all nonzero $x \in (\ker \tilde{R}_\lambda^{\frac{1}{2}} \tilde{P})^{\perp}$. Using the standard property of the adjoint and the definition of $\bar{P}$ we get that the above holds if and only if

$$\langle \bar{P} x, (\bar{P} Q \bar{P}^*) \bar{P} x \rangle < \langle \bar{P} x, \bar{P} x \rangle$$

for all nonzero *Euclidean* vectors $x \in (\ker \bar{P})^{\perp}$, which is equivalent to the inequality in the claim. ■

To check this condition we must compute $\check{C}^* \check{C}$ and $\tilde{P}^* \tilde{R}_\lambda \tilde{P}$. These are, by now, standard computations associated with the $\mathcal{H}_\infty$ sampled-data problem; see Appendix D for explicit formulae.

The next result constructs a matrix test to determine whether a particular scalar is an eigenvalue of our sampled-data operator.

Proposition 6.22 *Suppose that $\lambda > 0$ and is not in the spectrum of $\check{D}^* \check{D}$. Then λ is an eigenvalue of the operator $\check{M}^* \check{M}$ if and only if the matrix $I - Q \tilde{P}^* \tilde{R}_\lambda \tilde{P}$ is singular.*

This result differs from last mainly in that the quantity λ can be any scalar, that is not in the spectrum of $\check{D}^* \check{D}$. A similar and independent derivation appears in [69].

Proof The scalar λ is an eigenvalue if and only if there exists $\psi \in \mathcal{K}_2$ so that

$$(I\lambda - \check{M}^* \check{M}) \psi = 0.$$

This holds if and only if

$$(I - \tilde{R}_\lambda \tilde{P} Q \tilde{P}^*) \psi = 0,$$

where we have used the substitutions in (6.23) and (6.24). We can rewrite this statement as the existence of a complex vector x so that

$$(I - Q\tilde{P}^*\tilde{R}_\lambda\tilde{P})x = 0,$$

where $x = Q\tilde{P}^*\psi$ and

$$\psi = \tilde{R}_\lambda\tilde{P}x \tag{6.25}$$

■

This proof also tells us how to generate the eigenspace corresponding to the eigenvalue λ:

Corollary 6.23 *Suppose λ is not in* $\text{spec}(\check{D}^*\check{D})$. *If λ is an eigenvalue of $\check{M}^*\check{M}$, then*

$$\ker(I\lambda - \check{M}^*\check{M}) = \text{Im }(\tilde{R}_\lambda\tilde{P}V),$$

where V is any matrix satisfying $\text{Im}V = \ker(I - Q\tilde{P}^*\tilde{R}_\lambda\tilde{P})$.

Proof This follows directly from (6.25) in the proof of the above proposition. ■

The last three results provide a procedure to determine the largest eigenvalue of $\check{M}^*\check{M}$ to any desired degree of accuracy; check whether a scalar is an eigenvalue; and generate the eigenspace corresponding to an eigenvector.

Typically when constructing a map V, as above, we want $\tilde{R}_\lambda\tilde{P}V$ to be isometric, and this is easily accomplished: let U_0 be any injective matrix whose image is $\ker(I - Q\tilde{P}^*\tilde{R}_\lambda\tilde{P})$; then

$$\ker(I\lambda - \check{M}^*\check{M}) = \text{Im }(\tilde{R}_\lambda\tilde{P}U_0).$$

We require an isometric version of this mapping: namely to find a matrix U_1 so that

$$U_1^*U_0^*\tilde{P}^*\tilde{R}_\lambda^2\tilde{P}U_0U_1 = I.$$

Since the rank of U_0 and $\tilde{R}_\lambda\tilde{P}U_0$ are equal, and U_0 is injective, any invertible matrix U_1 satisfying

$$U_1U_1^* = (U_0^*\tilde{P}^*\tilde{R}_\lambda^2\tilde{P}U_0)^{-1}$$

fulfills our requirements by setting $V = U_0U_1$. A state space solution for $\tilde{P}^*\tilde{R}_\lambda^2\tilde{P}$ is given in Appendix D.

We now combine our results to get a version of Corollary 6.20 that can be verified explicitly. Before doing this define the operators E_k on $\mathcal{K}_2^{m+r}$ of the form

$$E_k = \text{diag}(0,\ldots,0,I,0,\ldots,0), \tag{6.26}$$

to be those that have an $m_k \times m_k$ identity block in the kth block as defined by the structure of $\mathcal{D}_{\Delta_{rp}}$. Thus, if $D = \text{diag}(d_1 I,\ldots,d_{d+1}) \in \mathcal{D}_{\Delta_{rp}}$ then $D = \sum_{k=1}^{d+1} d_k E_k$. Also, note that the operators E_k form a basis for $T_I(\mathcal{D}_{\Delta_{rp}})$, which is a fact used to prove the next theorem.

Theorem 6.24 *Suppose*

(i) λ *is not in* $spec(\check{D}^*\check{D})$, *where* $\lambda = \lambda_{max}(\check{M}^*\check{M})$

(ii) V *is a matrix satisfying* $\text{Im}V = \ker(I - Q\tilde{P}^*\tilde{R}_\lambda\tilde{P})$, *and* $U := \tilde{R}_\lambda\tilde{P}V$.

Then $\|\check{M}\| > \inf_{D\in\mathcal{D}_{\Delta_{rp}}} \|D^{\frac{1}{2}}\check{M}D^{-\frac{1}{2}}\|$ *if and only if there exist scalars* $y_k \in \mathbb{R}$ *so that*

$$\sum_{k=0}^{d+1} y_k(U^*\check{M}^*E_k\check{M}U - \lambda U^*E_kU) < 0.$$

Proof By Corollary 6.23 we have that $\ker(I\lambda - \check{M}^*\check{M}) = \text{Im}\,\tilde{R}_\lambda\tilde{P}V = \text{Im}\,U$. Thus, invoking Corollary 6.20 we have $\|\check{M}\| > \inf_{D\in\mathcal{D}_{\Delta_{rp}}} \|D^{\frac{1}{2}}\check{M}D^{-\frac{1}{2}}\|$ if and only if there exists $Y = \text{diag}(y_1,\ldots,y_{d+1}) \in T_I(\mathcal{D}_{\Delta_{rp}})$ so that

$$\begin{aligned} 0 > U^*(\check{M}^*Y\check{M} - \lambda Y)U &= U^*(\check{M}^*(\sum_{k=1}^{d+1} y_kE_k)\check{M} - \lambda\sum_{k=1}^{d+1} y_kE_k)U \\ &= \sum_{k=0}^{d+1} y_k(U^*\check{M}^*E_k\check{M}U - \lambda U^*E_kU) < 0. \quad \blacksquare \end{aligned}$$

This result reduces the optimality of a scale to a linear matrix inequality (LMI) in terms of the matrices $U^*\check{M}^*E_k\check{M}U - \lambda U^*E_kU$; formulae to compute these matrices can be found in Appendix D.

This LMI is of the standard form associated with the matrix structured singular value upper bound. The LMI problem can be answered using various techniques; see for example [12]. These correspond to a $Y \in T_I(\mathcal{D}_{\Delta_{rp}})$ which gives a direction of descent. Hence, we can compute the basic objects required for a descent algorithm. A constructive algorithm for computing optimal scalings will be discussed next.

6.3 Example Algorithm

Our intent here is to illustrate the results with an example. In the first subsection we apply our results to developing a cutting plane algorithm for the robust performance condition introduced in Subsection 6.1.2 using the tools developed in Section 6.2. The second subsection applies this algorithm to a numerical example.

6.3.1 A Cutting Plane Approach

We focus on the decision problem introduced in 6.2.1 which is, given a $\delta > 0$, does there exist a $D \in \mathcal{D}_{\Delta_{rp}}$ so that

$$\check{M}^* D \check{M} - \delta D \leq 0? \tag{6.27}$$

By Proposition 6.13 this immediately gives us an upper bound of δ on $\inf_{D \in \mathcal{D}_{\Delta_{rp}}} \|D^{\frac{1}{2}} \check{M} D^{-\frac{1}{2}}\|$. The approach we adopt is to formulate this problem as a convex optimization for which we develop a cutting plane algorithm. We start by giving a concise description of the main idea behind cutting plane methods, and then proceed to constructing our particular algorithm. We refer the reader to [40] or [13] for further information on cutting plane techniques, and remark that in [14] they are used for solving LMI problems.

In general, cutting plane methods are applied to optimization problems with the following format: given a family of compact convex sets

$\mathcal{W}_1, \ldots, \mathcal{W}_q$ in $\mathbb{R}^p$, and a vector $c \in \mathbb{R}^p$ solve

$$\begin{array}{ll} \text{minimize} & \langle c, x \rangle \\ \text{subject to} & x \in \cap_{l=1}^{q} \mathcal{W}_l \end{array} \tag{6.28}$$

Note that no feasible solution to this problem may exist if the sets $\mathcal{W}_l$ have no common intersection.

A cutting plane algorithm for (6.28) proceeds iteratively in the following way: start, for some n, with a compact convex set $\mathcal{P}_n$ which contains the intersection $\cap_{l=1}^{q} \mathcal{W}_l$, and is defined by the set of elements $x \in \mathbb{R}^p$ that satisfy the list of linear inequalities

$$\begin{array}{rcl} \langle a_1, x \rangle & \leq & b_1 \\ & \vdots & \\ \langle a_n, x \rangle & \leq & b_n, \end{array}$$

for fixed vectors a_l in $\mathbb{R}^p$ and real scalars b_l. Then solve the less constrained problem

$$\begin{array}{ll} \text{minimize} & \langle c, x \rangle \\ \text{subject to} & x \in \mathcal{P}_n \end{array}$$

which is a linear program. If no feasible solution exists, then (6.28) does not have a feasible solution and the algorithm terminates. Otherwise, let x^0 be the optimal solution to the linear program. If x^0 is in $\cap_{l=1}^{q} \mathcal{W}_l$, then it is the optimal solution to (6.28) and no further iterations are necessary. However, when x^0 is not a feasible solution, a vector a_{n+1} and scalar b_{n+1} are found so that

$$\begin{array}{rcll} \langle a_{n+1}, x \rangle & \leq & b_{n+1} & \text{for all } x \in \cap_{l=1}^{q} \mathcal{W}_l. \\ \langle a_{n+1}, x^0 \rangle & > & b_{n+1}. & \end{array} \tag{6.29}$$

Namely, this defines a condition which separates x^0 from the feasible set $\cap_{l=1}^{q} \mathcal{W}_l$. The constraint defined by (6.29) is added to the list of constraints defining $\mathcal{P}_n$ to obtain an updated set $\mathcal{P}_{n+1}$, which now contains $\cap_{l=1}^{q} \mathcal{W}_l$, and is a strict subset of $\mathcal{P}_n$. The procedure is then repeated.

Now, we move on to developing a cutting plane algorithm specific to the question posed in (6.27). Consider the following convex optimization in $\mathbb{R}^{d+2}$:

$$\begin{array}{lll} \text{minimize} & -d_0 & \\ \text{subject to} & (i) & Id_0 + \sum_{k=1}^{d+1} d_k(\check{M}^* E_k \check{M} - \delta E_k) \leq 0 \quad \text{in } \mathfrak{L}(\mathcal{K}_2) \\ & (ii) & d_0 \geq 0 \text{ and } d_k > 0 \text{ for } 1 \leq k \leq d+1, \end{array} \tag{6.30}$$

where the operators E_k are defined in (6.26). The relationship of this optimization to the problem in (6.27) is seen by recalling that if $D \in \mathcal{D}_{\Delta_{rp}}$ and $D =: \text{diag}(d_1 I, \ldots, d_{d+1} I)$, then $D = \sum_{k=1}^{d+1} d_k E_k$. Therefore, (6.30) has a feasible solution if and only if (6.27) does.

It is clear that the constraints above are convex, but they do not define compact sets in $\mathbb{R}^{d+2}$. Henceforth, we assume that additional constraints of lower and upper bounds for the variables d_k are imposed so that they lie in compact sets; such numerical assumptions must always be made in order to develop a practical algorithm that addresses (6.27). With this additional assumption the optimization in (6.30) is exactly of the form in (6.28).

Our main goal is to generate a sequence of sets $\mathcal{P}_n$ for this problem. State space formulae for all the objects we require in the construction can be found in Appendix D. Start with a compact set $\mathcal{P}_n$ of the form in (6.29), for some value of n, which contains all the points satisfying the constraints of (6.30). And let $(d_0^0, \ldots, d_{d+1}^0)$ be the solution to the linear optimization

$$\begin{array}{ll} \text{minimize} & -d_0 \\ \text{subject to} & (d_0, \ldots, d_{d+1}) \in \mathcal{P}_n \\ & d_0 \geq 0 \text{ and } d_k > 0 \text{ for } 1 \leq k \leq d+1. \end{array} \tag{6.31}$$

Then the first step is to determine whether $(d_0^0, \ldots, d_{d+1}^0)$ is a feasible solution to our optimization problem. A *necessary* condition for this is that

$$\sum_{k=1}^{d+1} d_k^0(\check{M}^* E_k \check{M} - \delta E_k) \leq 0,$$

since condition (i) in (6.30) must be met and $d_0^0 \geq 0$. The condition is equivalent to (6.27) with D set to $D_0 := \operatorname{diag}(d_1^0 I, \ldots, d_{d+1}^0 I)$. Now, by Lemma 6.15 we have that this holds if and only if $\lambda_{\max}(D_0^{-\frac{1}{2}} M^* D_0 M D_0^{-\frac{1}{2}}) \leq \delta$.

Step 1: Compute $\lambda_{\max}(D_0^{-\frac{1}{2}} M^* D_0 M D_0^{-\frac{1}{2}})$ using Propositions 6.21 and 6.22.

If the resulting eigenvalue is less than or equal to δ then we have satisfied (6.27) and we stop. Otherwise, (6.27) and the constraint in (6.30) are violated, and we must construct a new constraint. To accomplish this we find the eigenspace associated with
$\bar{\lambda} := \lambda_{\max}(D_0^{-\frac{1}{2}} M^* D_0 M D_0^{-\frac{1}{2}})$.

Step 2: Use Corollary 6.23 to obtain an injective map $U_0 : \mathbb{C}^u \to \mathcal{K}_2$, for some dimension u, that satisfies $\ker(I\bar{\lambda} - D_0^{-\frac{1}{2}} M^* D_0 M D_0^{-\frac{1}{2}}) = \operatorname{Im} U_0$.

From the definition of U_0 we have that

$$U_0^* (D_0^{-\frac{1}{2}} (M^* D_0 M - I\bar{\lambda} D_0) D_0^{-\frac{1}{2}} U_0 = 0.$$

So,

$$U_0^* D_0^{-\frac{1}{2}} (M^* D_0 M - \delta D_0) D_0^{-\frac{1}{2}} U_0 = (\bar{\lambda} - \delta) U_0^* U_0 > 0, \tag{6.32}$$

where the RHS follows because U_0 is injective and $\bar{\lambda} > \delta$. Now, choose an invertible matrix V, in the manner described after Corollary 6.23, so that $V^* U_0^* D^{-1} U_0 V = I$. Then set $U := D_0^{-\frac{1}{2}} U_0 V$. We have that any feasible solution to (6.30) necessarily satisfies

$$I d_0 + \sum_{k=1}^{d+1} d_k U^* (\check{M}^* E_k \check{M} - \delta E_k) U \leq 0 \quad \text{in } \mathbb{C}^{u \times u} \tag{6.33}$$

where we have premultiplied and postmultiplied the linear *operator* inequality in (i) by U^* and U, respectively, to obtain a linear *matrix* inequality. By (6.32) it follows that

$$I d_0^0 + \sum_{k=1}^{d+1} d_k^0 U^* (\check{M}^* E_k \check{M} - \delta E_k) U > 0.$$

We have therefore succeeded in separating the point $(d_0^0, \ldots, d_{d+1}^0)$, obtained as the solution to the linear optimization (6.31), from the feasible points of (6.30).

We can obtain a single linear constraint from (6.33): postmultiply and premultiply by any unit vector in $\mathbb{C}^u$ and its transpose to get a constraint of the form

$$\langle a_{n+1}, \underline{d} \rangle \leq b_{n+1},$$

where $\underline{d} = (d_0, \ldots, d_{d+1})$, a_{n+1} is an appropriate vector and b_{n+1} a corresponding scalar. This constraint is now included with those that define $\mathcal{P}_n$ to get the set $\mathcal{P}_{n+1}$.

Step 3: Update from n to $n+1$ and solve the linear optimization in (6.31).

If no feasible solution exists then (6.27) cannot be solved. Otherwise we return to Step 1 with this updated candidate solution. This completes the procedure.

We further remark that the above algorithm can be modified to use the LMI constraints generated by (6.33) directly, thus producing a sequence of linear optimizations with LMI constraints. Methods exist for solving such problems; see for example [12] for a synopsis of available LMI optimization techniques.

6.3.2 Numerical Example

We now consider a simple numerical example in order to demonstrate the algorithm above and the robustness results of the chapter; the example is intentionally chosen to be as simple as possible. The system we concentrate on is shown in Figure 6.2, which fits the general robust performance structure depicted in Figure 6.1. In the configuration we have a single perturbation $\boldsymbol{\Delta}$ which has scalar valued inputs and outputs. The signals w and z are also scalar valued, and will be in $\mathcal{L}_2$. The controller is a unity feedback with two measured inputs and two control channels. In our

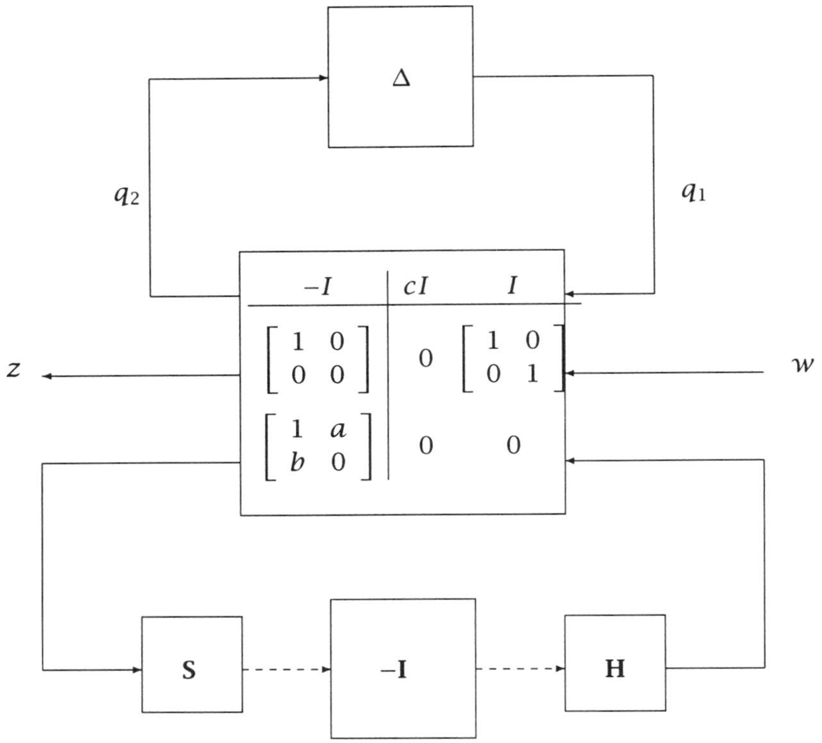

Figure 6.2: System Configuration

analysis we focus on robust performance of this system to perturbations Δ that are quasi-periodic. The parameters of the system are $a = 7$, $b = 0.4$, $c = 0.1$, and the sampling rate used is $h = 0.7$. This system is unstable for h greater than 0.8.

Along with the results for this sampled-data setup we also show the analogous calculations for the discretized sampled-data system. Namely, the system M_d with inputs and outputs in ℓ_2, obtained from Figure 6.1, by defining $M_d :=$ **SMH**. We denote its discrete time transfer function by $M_d(z)$.

We first calculate the norm of the sampled-data transfer function $\|\check{M}(e^{j\omega})\|$ with respect to the variable ω. This function is shown by the solid line in Figure 6.3. The values $\|\check{M}(e^{j\omega})\|$ at each frequency have been calculated, iteratively, using the bisection method based on Propositions 6.21 and 6.22. Also shown on the plot is the maximum singular value $\bar{\sigma}(M_d(e^{j\omega}))$ of the discrete time transfer function $M_d(e^{j\omega})$. We see that everywhere this transfer function norm lies below the norm of our sampled-data transfer function.

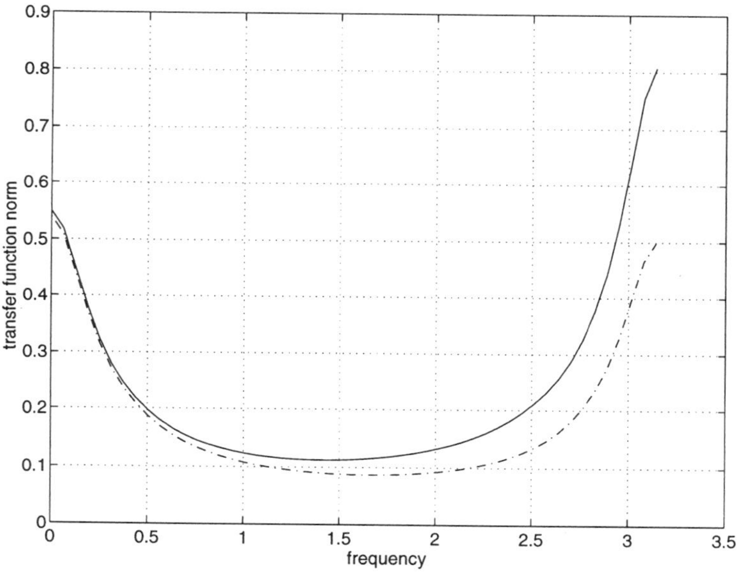

Figure 6.3: Plots of $\|\check{M}(e^{j\omega})\|$ and $\bar{\sigma}(M_d(e^{j\omega}))$

To address the robust performance of sampled-data system, we calculate the quantity $\inf_{D\in\mathcal{D}_{\Delta_{rp}}} \|D^{\frac{1}{2}}\check{M}D^{-\frac{1}{2}}\|$ at every frequency ω. For this particular problem the set

$$\mathcal{D}_{\Delta_{rp}} = \{\text{diag}(d_1 I, d_2 I) \in \mathfrak{L}(\mathcal{K}_2^2) : d_k > 0\},$$

since the signals w, z, q_1 and q_2 are scalars.

We minimize the weighted norm $\|D^{\frac{1}{2}}\check{M}(e^{j\omega})D^{-\frac{1}{2}}\|$ using the cutting plane approach developed previously. At each frequency ω a scaling $D_\omega \in$

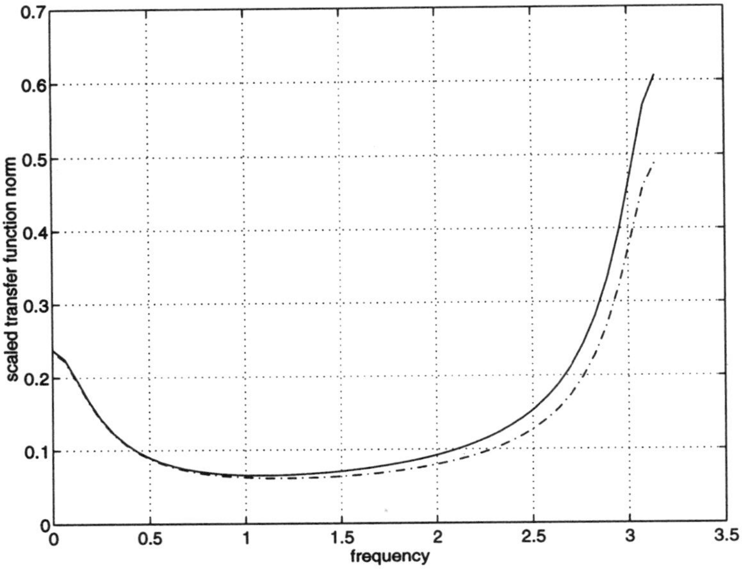

Figure 6.4: Graphs of $\|D^{\frac{1}{2}}\check{M}(e^{j\omega})D^{-\frac{1}{2}}\|$ and $\bar{\sigma}(D^{\frac{1}{2}}M_d(e^{j\omega})D^{-\frac{1}{2}})$

$\mathcal{D}_{\Delta_{rp}}$ is found so that

$$\begin{aligned}(1-\epsilon_{tol})\|D_\omega^{\frac{1}{2}}\check{M}(e^{j\omega})D_\omega^{-\frac{1}{2}}\| &\le \inf_{D\in\mathcal{D}_{\Delta_{rp}}} \|D^{\frac{1}{2}}\check{M}(e^{j\omega})D^{-\frac{1}{2}}\| \\ &\le (1+\epsilon_{tol})\|D_\omega^{\frac{1}{2}}\check{M}(e^{j\omega})D_\omega^{-\frac{1}{2}}\|\end{aligned}$$

is satisfied. The graph resulting from the optimization carried out with $\epsilon_{tol} = 0.005$ is shown by the solid curve in Figure 6.4.

The broken line in the figure is the graph of the pointwise minimization of $\bar{\sigma}(DM_d(e^{j\omega})D^{-1})$ over the nonsingular matrices $\text{diag}(d_1, d_2)$; this is the standard structured singular value upper bound for the matrix described in Section 2.6. Both graphs lie significantly below their corresponding transfer functions in Figure 6.3.

Pertaining to the minimization $\|D^{\frac{1}{2}}\check{M}D^{-\frac{1}{2}}\|$ using the cutting plane algorithm, we show plots of $(d_2/d_1)^{\frac{1}{2}}$ where $D_\omega = \text{diag}(d_1 I, d_2 I)$ for $\epsilon_{tol} = \{0.005, 10^{-4}\}$. The solid curve in Figure 6.5 corresponds to the optimiza-

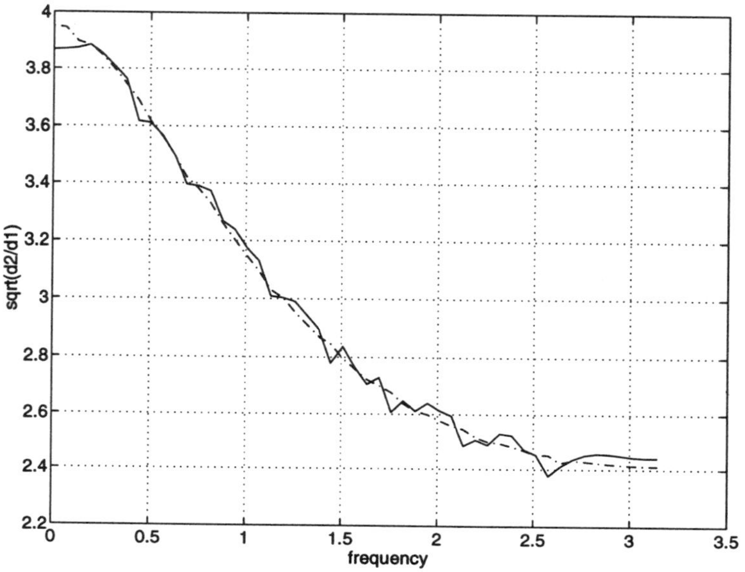

Figure 6.5: Plots of $(d_2/d_1)^{\frac{1}{2}}$ with $\epsilon_{tol} = \{0.005, 10^{-4}\}$

tion carried out with $\epsilon_{tol} = 0.005$. We see this curve is not smooth; this is because although the objective function $\|D^{\frac{1}{2}}\check{M}(e^{j\omega})D^{-\frac{1}{2}}\|$ has been calculated to a tolerance of 0.005, this does not impose the same tolerance on the scaling $D_{\omega}^{\frac{1}{2}}$. The dotted curve of Figure 6.5 shows the result of the optimization when $\epsilon_{tol} = 10^{-4}$, which is sufficiently small to produce a smooth scaling curve.

Comparing the graphs of $\|\check{M}(e^{j\omega})\|$ and $\|D_{\omega}^{\frac{1}{2}}\check{M}(e^{j\omega})D_{\omega}^{-\frac{1}{2}}\|$, we see the maximum of approximately 0.8 in Figure 6.3 has been reduced to near 0.6 in Figure 6.4. Invoking Theorem 6.6 we can then be guaranteed that, for every $r < 1/0.6$, there exists $\nu > 0$, so that the gain $\|w \mapsto z\|_{\mathcal{L}_2 \to \mathcal{L}_2}$ will be less than or equal to 0.6 for all perturbations $\boldsymbol{\Delta}$ in the open unit ball of $\mathfrak{X}_{QP}(\nu)$ with radius r. Furthermore, by Theorem 6.5, if $r > 1/0.6$ then for each $\nu > 0$ there exists $\boldsymbol{\Delta}$ in the open unit ball of $\mathfrak{X}_{QP}(\nu)$ with radius r such that $\|w \mapsto z\|_{\mathcal{L}_2 \to \mathcal{L}_2} > 0.6$. We remark that in Figure 6.4 the gap is greater than 0.1 between the maximum of $\|D_\omega^{\frac{1}{2}}\check{M}(e^{j\omega})D_\omega^{-\frac{1}{2}}\|$ and the maximum of $\bar{\sigma}(DM_d(e^{j\omega})D^{-1})$. The maximum of $\bar{\sigma}(DM_d(e^{j\omega})D^{-1})$ is one traditional method of assessing robust performance of Figure 6.2.

Next we investigate an upper bound for $\inf_{D\in\mathcal{D}_{\Delta rp}} \|D^{\frac{1}{2}}\check{M}D^{-\frac{1}{2}}\|$ which complements the current one, and has significant computational advantages.

6.4 Minimizing the Scaled Hilbert-Schmidt Norm

Up until now we have been concerned with minimizing the scaled induced norm $\|D^{\frac{1}{2}}\check{M}D^{-\frac{1}{2}}\|$. In this section we concentrate on a simpler problem: that of minimizing the Hilbert-Schmidt norm $\|D^{\frac{1}{2}}\check{M}D^{-\frac{1}{2}}\|_{\text{HS}}$ over the same scaling set $\mathcal{D}_{\Delta_{rp}}$.

In the matrix case of computing the upper bound of the structured singular value, see for example [49][72], minimizing the scaled Frobenius norm is an effective first step in minimizing the scaled maximum singular value of a matrix. The problem we examine here is the analog of this in infinite dimensions. We show that minimizing the scaled Hilbert-Schmidt norm of $\check{M}(e^{j\omega_0})$ is equivalent to minimizing the scaled Frobenius norm of a particular matrix. Hence, standard numerical algorithms can be applied to this problem. We remark that the Hilbert-Schmidt norm has also been used in a sampled-data context in [6] to solve a version of the $\mathcal{H}_2$-optimal sampled-data problem.

The section is divided into two short parts: we introduce the Hilbert-Schmidt norm, which is subsequently applied.

6.4.1 The Hilbert-Schmidt Norm

The material here is standard; see for example [71]. An operator Q on a Hilbert space $\mathcal{H}$ is by definition a Hilbert-Schmidt operator if there is an orthonormal basis $\{\phi_k\}_{k=1}^{\infty}$ for $\mathcal{H}$ so that

$$\|Q\|_{\text{HS}} := (\sum_{k=1}^{\infty} \|Q\phi_k\|_{\mathcal{H}}^2)^{\frac{1}{2}} < \infty, \tag{6.34}$$

which then defines the Hilbert-Schmidt norm when finite. It is easy to verify that this definition is basis independent. Furthermore, *all* Hilbert-Schmidt operators are compact.

We are particularly interested in integral kernel operators. Suppose Q is an operator on $\mathcal{K}_2^m$ and is defined by the integral equation

$$(Qw)(t) := \int_0^h Q(t,\tau)w(\tau)\,d\tau, \tag{6.35}$$

where $Q(\cdot,\cdot)$ is a bounded matrix valued function on $[0,h]\times[0,h]$, and w is some element of $\mathcal{K}_2^m$. It is a fact that such an operator is Hilbert-Schmidt.

The definition for the Hilbert-Schmidt norm given in (6.34) is not particularly practical from a computational point of view, and we now develop a more convenient, and well-known, form for integral operators. Let $\{\phi_k\}_{k=1}^{\infty}$ be an orthonormal basis for the scalar functions $\mathcal{K}_2^1$, and define $\{e_l\}_{l=1}^{m}$ to be the standard basis for $\mathbb{C}^m$. Then for any integral operator Q as defined above we have that

$$\|Q\|_{\text{HS}}{}^2 = \sum_{l=1}^{m}\sum_{k=1}^{\infty} \|Qe_l\phi_k\|_{\mathcal{K}_2}^2,$$

since $e_l\phi_k$ is a basis for $\mathcal{K}_2^m$. Fix k and l, and we see that

$$\|Qe_l\phi_k\|_{\mathcal{K}_2}^2 = \int_0^h |(q_l^t,\phi_k)|_2^2 dt,$$

where $q_l^t := Q(t,\tau)e_l$ and $(q_l^t,\phi_k) = \int_0^h Q(t,\tau)e_l\phi_k(\tau)d\tau$, a vector of inner products. Hence, fixing $l \geq 0$ at some value we get

$$\begin{aligned}\sum_{k=1}^{\infty} \|Qe_l\phi_k\|_{\mathcal{K}_2}^2 &= \sum_{k=1}^{\infty}\int_0^h |(q_l^t,\phi_k)|_2^2 dt \\ &= \int_0^h (\sum_{k=1}^{\infty} |(q_l^t,\phi_k)|_2^2) dt.\end{aligned}$$

Then by Parseval's theorem, since $\{\phi_k\}_{k=1}^{\infty}$ is an orthonormal basis, we have for each $t \geq 0$ that $\sum_{k=0}^{\infty} |\langle q_l^t, \phi_k \rangle|_2^2 = \int_0^h |q_l^t(\tau)|_2^2 \, d\tau = \|(q_l^t)^* q_l^t\|_{\mathcal{K}_2}^2$. So,

$$\begin{aligned}\sum_{k=1}^{\infty} \|Qe_l\phi_k\|_{\mathcal{K}_2}^2 &= \int_0^h \|(q_l^t)^* q_l^t\|_{\mathcal{K}_2}^2 \, dt \\ &= \int_0^h \int_0^h e_l^* Q^*(t,\tau) Q(t,\tau) e_l \, d\tau \, dt.\end{aligned}$$

Summing all m terms we get

$$\|Q\|_{\text{HS}}{}^2 = \sum_{l=1}^{m} \sum_{k=1}^{\infty} \|Qe_l\phi_k\|_{\mathcal{K}_2}^2 = \int_0^h \int_0^h \operatorname{tr}(\, Q^*(t,\tau) Q(t,\tau)\,) \, d\tau \, dt, \qquad (6.36)$$

where $\operatorname{tr}(\cdot)$ denotes the trace of the argument matrix. For integral operators, we have therefore converted the abstract definition of the Hilbert-Schmidt norm in (6.34) to the above integral form.

For reference in the next section define the standard Frobenius norm, $|\cdot|_{\text{F}}$, of a matrix X by

$$|X|_{\text{F}}^2 := \operatorname{tr}(X^* X).$$

With these definitions we move to the next subsection which applies them to our sampled-data operator $\check{M}$.

6.4.2 Scaling the Hilbert-Schmidt Norm and Osborne's Method

To start, we have a result which establishes that $\check{M}(e^{j\omega_0})$ is an integral operator for each $\omega_0 \in (-\pi, \pi]$ and is therefore Hilbert-Schmidt. *Throughout* we fix $\omega_0 \in (-\pi, \pi]$ and write $\check{M}$ instead of $\check{M}(e^{j\omega_0})$.

Proposition 6.25 *The operator $\check{M}(e^{j\omega_0})$ on $\mathcal{K}_2$ is Hilbert-Schmidt.*

Proof It is straightforward to verify from (3.5) that $\check{M}(e^{j\omega_0})$ can be defined from the integral kernel

$$M(t,\tau) := \begin{cases} C_{\check{C}}\, e^{A_{\check{C}}t}\, B_{\check{C}} e^{j\omega_0}(I - e^{j\omega_0}A_d)^{-1} C_{\check{B}} e^{A(h-\tau)} B_1 & \\ \qquad + C_1 e^{A(t-\tau)} B_1 & \tau \in [0, t] \\ & \\ C_{\check{C}}\, e^{A_{\check{C}}t}\, B_{\check{C}} e^{j\omega_0}(I - e^{j\omega_0}A_d)^{-1} C_{\check{B}} e^{A(h-\tau)} B_1 & \tau \in (t, h), \end{cases} \tag{6.37}$$

where the matrices $C_{\check{C}}$, $A_{\check{C}}$, $B_{\check{C}}$ and $A_{\check{C}}$ are given in Appendix F. Hence, $\check{M}$ is Hilbert-Schmidt. ■

The main motivation for using the Hilbert-Schmidt norm in this context stems from the following property: if Q is a Hilbert-Schmidt operator then

$$\|Q\|_{\text{HS}}^2 = \sum_{k=1}^{\infty} \lambda_k,$$

where the λ_k are the eigenvalues of Q^*Q. This property is easy to see from the definition of the norm in (6.34), and the fact that a self-adjoint compact operator is diagonalizable. The implication of this is that, for any D in our scaling set $\mathcal{D}_{\Delta_{rp}}$, we have that

$$\|D^{\frac{1}{2}}\check{M}D^{-\frac{1}{2}}\|^2 = \lambda_{\max}(D^{-\frac{1}{2}}\check{M}^*D\check{M}D^{-\frac{1}{2}}) \leq \|D^{\frac{1}{2}}\check{M}D^{-\frac{1}{2}}\|_{\text{HS}}^2.$$

In particular, if the eigenvalues of $D^{-\frac{1}{2}}\check{M}^*D\check{M}D^{-\frac{1}{2}}$ fall off rapidly then the LHS and RHS above will be close. Even when this is not the case, the $D \in \mathcal{D}_{\Delta_{rp}}$ obtained can frequently be near the optimal scale for the LHS.

We now aim to show how the minimization of $\|D^{\frac{1}{2}}\check{M}D^{-\frac{1}{2}}\|_{\text{HS}}$ can be reduced to a well-known optimization problem. Define the $d+1 \times d+1$ matrix X with entries

$$(X)_{kl} := \|E_k\check{M}E_l\|_{\text{HS}}, \tag{6.38}$$

where the set of matrices E_k are, as defined in (6.26). Explicit expressions for the scalars $\|E_k\check{M}E_l\|_{\text{HS}}$ can be found in Appendix F, and have been obtained using the integral kernel in the proof of Proposition 6.25. Also, define the related set of matrices

$$\mathcal{Y} := \{Y \in \mathbb{R}^{d+1\times d+1} : Y := \text{diag}(y_1, \ldots, y_{d+1}),\ y_k > 0\}.$$

We can now prove the main result of this section.

Theorem 6.26 *The following equality holds:*

$$\inf_{D \in \mathcal{D}_{rp}} \|D^{\frac{1}{2}} \check{M} D^{-\frac{1}{2}}\|_{HS} = \inf_{Y \in \mathcal{Y}} |YXY^{-1}|_F,$$

where X and $\mathcal{Y}$ are defined above.

Proof Let $D = \mathrm{diag}(d_1 I, \ldots, d_{d+1} I) \in \mathcal{D}_{\Delta_{rp}}$. From (6.36) and (6.37) we have that

$$\begin{aligned} \|D^{\frac{1}{2}} \check{M} D^{-\frac{1}{2}}\|_{\mathrm{HS}}{}^2 &= \int_0^h \int_0^h \mathrm{tr}\{D^{-\frac{1}{2}} M^*(t,\tau) D M(t,\tau) D^{-\frac{1}{2}}\}\, d\tau dt \\ &= \int_0^h \int_0^h \mathrm{tr}\{M^*(t,\tau) D M(t,\tau) D^{-1}\}\, d\tau dt. \end{aligned}$$

Substituting $D = \sum_{k=1}^{d+1} d_k E_k$ into the above we get

$$\begin{aligned} \|D\check{M}D^{-1}\|_{\mathrm{HS}}{}^2 &= \int_0^h \int_0^h \mathrm{tr}\{M^*(t,\tau)(\sum_{k=1}^{d+1} d_k E_k) M(t,\tau)(\sum_{l=1}^{d+1} d_l^{-1} E_l)\}\, d\tau dt \\ &= \sum_{k=1}^{d+1} \sum_{l=1}^{d+1} d_k d_l^{-1} \int_0^h \int_0^h \mathrm{tr}\{M^*(t,\tau) E_k M(t,\tau) E_l\}\, d\tau dt. \end{aligned}$$

Noting that $E_k^2 = E_k$ for all $1 \le k \le d+1$ we get the above is equal to

$$\begin{aligned} &\sum_{k=1}^{d+1} \sum_{l=1}^{d+1} d_k d_l^{-1} \int_0^h \int_0^h \mathrm{tr}\{E_l M^*(t,\tau) E_k M(t,\tau) E_l\}\, d\tau dt \\ &= \sum_{k=1}^{d+1} \sum_{l=1}^{d+1} d_k d_l^{-1} \|E_k \check{M} E_l\|_{\mathrm{HS}}{}^2 \\ &= |YXY^{-1}|_{\mathrm{F}}^2, \end{aligned}$$

where $Y^2 := \mathrm{diag}(d_1, \ldots, d_{d+1})$. The proof is then completed by starting with an element $Y \in \mathcal{Y}$ and reversing the above argument. ■

The theorem reduces the problem of minimizing the scaled Hilbert-Schmidt norm to a Frobenius norm optimization on matrices. This minimization can be solved by Osborne's method [46], which has been used extensively in the matrix structured singular value problem.

6.5 Summary

We have provided a detailed analysis of robust performance of sampled-data systems to periodic, quasi-periodic and arbitrary time-varying structured perturbations. Our results show that robustness to structured periodic uncertainty can be evaluated using the structured singular value of the nominal sampled-data transfer function. We have demonstrated an exact test for robust performance to a class of quasi-periodic structured perturbations, and also provided a necessary and sufficient condition for robust performance to arbitrary time-varying uncertainty.

Computation of the quasi-periodic and arbitrary time-varying robustness conditions have been shown to be quasi-convex optimization problems on Euclidean space. These optimization problems are similar to a type of LMI problem. We have presented several tools for explicitly evaluating various objects associated with the quasi-periodic condition, and have given an example cutting plane technique to check the condition. A direction for future work is the development of an efficient numerical algorithm, such as an interior point method, following [44], to compute this condition. Note: these computational tools for the quasi-periodic case can be readily extended, by a routine development, to also handle the condition for arbitrary time-varying uncertainty presented in Theorem 6.12.

The results of this chapter can be used in a practical robust synthesis procedure similar to the μ-synthesis iteration outlined in 5.3. Also of note is that by using the D-scales referred to in 5.3, it is possible to derive exact convex conditions for robust stability to quasi-LTI perturbations using the approach of 6.1.2. The resulting conditions are however infinite dimensional.

To complement the above quasi-periodic robustness condition, we have developed an additional condition in terms of the weighted Hilbert-Schmidt norm of the sampled-data transfer function. This condition has significant computational advantages in that it exactly reduces to a well-known matrix optimization problem for which efficient algorithms exist

[46]. These results also extend so that a Hilbert-Schmidt norm optimization can be derived which complements the arbitrary time-varying result in Theorem 6.12.

Appendix A

State space for $\tilde{M}$

Provided in this appendix is a derivation of the state space for the operator $\tilde{M}$ in (3.4). The derivation is based on that in [8]. To start, examine the operator on $\mathbf{M} = \mathcal{F}_l(\mathbf{G}, \mathbf{HK}_d\mathbf{S})$: recall that $(A_{K_d}, B_{K_d}, C_{K_d}, D_{K_d})$ is a minimal state space realization for $\mathbf{K}_d$, and

$$\hat{G}(s) = \left[\begin{array}{c|cc} A & B_1 & B_2 \\ \hline C_1 & 0 & D_{12} \\ C_2 & 0 & 0 \end{array}\right],$$

is the realization for $\mathbf{G}$ defined in (3.1). Given $w \in \mathcal{L}_2$ we solve $z = \mathcal{F}_l(\mathbf{G}, \mathbf{HK}_d\mathbf{S})w$ by the equations

$$\begin{aligned} \begin{bmatrix} z \\ y \end{bmatrix} &= \begin{bmatrix} \mathbf{G}_{11} & \mathbf{G}_{12} \\ \mathbf{G}_{21} & \mathbf{G}_{22} \end{bmatrix} \begin{bmatrix} w \\ u \end{bmatrix} \\ u &= \mathbf{HK}_d\mathbf{S}\, y. \end{aligned}$$

From the above state space realization, and the definitions of the sample and hold operators in (2.2), these equations can be realized by

$$\begin{aligned} \dot{x}_G &= Ax_G(t) + B_1 w(t) + B_2 u[k], \qquad x_G(0) = 0 \\ z(t) &= C_1 x_G(t) + D_{12} u[k] \\ y(t) &= C_2 x_G(t) \end{aligned}$$

$$\begin{aligned} x_{K_d}[k+1] &= A_{K_d} x_{K_d}[k] + B_{K_d} y(kh), \qquad x_{K_d}[0] = 0 \\ u[k] &= C_{K_d} x_{K_d}[k] + D_{K_d} y(kh), \end{aligned}$$

for $t \geq 0$ and k so that $t \in [kh, (k+1)h)$. From these equations it is straightforward to show that the following equations are satisfied:

$$\begin{aligned} x_G((k+1)h) &= e^{Ah} x_G(kh) + \int_0^h e^{A(h-\eta)} B_1 w(kh+\eta)\, d\eta \\ &\quad + \int_0^h e^{A\eta}\, d\eta B_2\, u[k] \\ z(\tau + kh) &= C_1 e^{A\tau} x_G(kh) + C_1 \int_0^\tau e^{A(\tau-\eta)} B_1 w(kh+\eta)\, d\eta \\ &\quad + (C_1 \int_0^\tau e^{A\eta}\, d\eta B_2 + D_{12})\, u[k], \end{aligned}$$

for any $\tau \in [0, h)$ and integer $k \geq 0$. We now set $\tilde{w} = Ww$ and $\tilde{z} = Wz$, where W is the lifting operator defined in (2.4). With these new signals defined, routine manipulations show that

$$\begin{aligned} \begin{bmatrix} x_G((k+1)h) \\ x_{K_d}[k+1] \end{bmatrix} &= A_d \begin{bmatrix} x_G(kh) \\ x_{K_d}[k] \end{bmatrix} + \check{B}\tilde{w}[k]; \quad \begin{bmatrix} x_G(0) \\ x_{K_d}[0] \end{bmatrix} = 0 \\ \tilde{z}[k] &= \check{C} \begin{bmatrix} x_G(kh) \\ x_{K_d}[k] \end{bmatrix} + \check{D}\tilde{w}[k] \end{aligned} \tag{A.1}$$

where $A_d \in \mathbb{C}^{\tilde{n}\times\tilde{n}}$, $\check{B} : \mathcal{K}_2 \to \mathbb{C}^{\tilde{n}}$, $\check{C} : \mathbb{C}^{\tilde{n}} \to \mathcal{K}_2$, and $\check{D} : \mathcal{K}_2 \to \mathcal{K}_2$ are provided below.

$$\begin{aligned} A_d &= \begin{bmatrix} e^{Ah} + \int_0^h e^{A(h-\eta)}\, d\eta B_2 D_{K_d} C_2 & \int_0^h e^{A(h-\eta)}\, d\eta B_2 C_{K_d} \\ B_{K_d} C_2 & A_{K_d} \end{bmatrix} \\ \check{B}\psi &= \begin{bmatrix} \int_0^h e^{A(h-\eta)} B_1 \psi(\eta)\, d\eta \\ 0 \end{bmatrix} \\ (\check{C}x)(\tau) &= \begin{bmatrix} C_1 e^{A\tau} & C_1 \int_0^\tau e^{A(\tau-\eta)}\, d\eta B_2 + D_{12} \end{bmatrix} \begin{bmatrix} I & 0 \\ D_{K_d} C_2 & C_{K_d} \end{bmatrix} x \\ (\check{D}\psi)(\tau) &= C_1 \int_0^\tau e^{A(\tau-\eta)} B_1 \psi(\eta)\, d\eta. \end{aligned} \tag{A.2}$$

The operator $\tilde{M} = W\mathcal{F}_l(\mathbf{G}, \mathbf{HK}_d\mathbf{S})W^{-1}$ is then given by (A.1).

The matrices $(\tilde{B})_k$, $(\tilde{C})_l$ and $(\tilde{D})_{lk}$ that form the frequency response $M(e^{j\omega_0})$, defined in (4.8), are obtained from the basis $\{\psi_k\}_{k=0}^{\infty}$ in (4.3) and are:

$$
\begin{aligned}
(\tilde{B})_k &= h^{-\frac{1}{2}} \begin{bmatrix} \int_0^h e^{A(h-\eta)} e^{j\theta_k \eta)}\, d\eta B_1 \\ 0 \end{bmatrix} \\
(\tilde{C})_l &= \begin{bmatrix} C_1 \int_0^h e^{A\tau} e^{-j\theta_l \tau)}\, d\tau & \int_0^h (C_1 \int_0^\tau e^{A(\tau-\eta)}\, d\eta B_2 + D_{12}) e^{-j\theta_l \tau}\, d\tau \end{bmatrix} \cdot \\
& \qquad h^{-\frac{1}{2}} \begin{bmatrix} I & 0 \\ D_{K_d} C_2 & C_{K_d} \end{bmatrix} \\
(\tilde{D})_{lk} &= h^{-1} C_1 \int_0^h e^{-j\theta_l \tau} \int_0^\tau e^{A(\tau-\eta)} e^{j\theta_k \eta}\, d\eta\, d\tau B_1,
\end{aligned}
$$

where $\theta_k = \frac{2\pi\nu_k - \omega_0}{h}$ and ν_k is the sequence $\{0, 1, -1, 2, -2, \ldots\}$.

Using the following formula, where E is a square matrix,

$$
\int_0^h e^{E\eta}\, d\eta = [I \;\; 0] \exp\left(\begin{bmatrix} E & I \\ 0 & 0 \end{bmatrix} h \right) \begin{bmatrix} 0 \\ I \end{bmatrix},
$$

and the above expressions for $(\tilde{B})_k$, $(\tilde{C})_l$ and $(\tilde{D})_{lk}$ can be rewritten as

$$
\begin{aligned}
(\tilde{B})_k &= h^{-\frac{1}{2}} \begin{bmatrix} I \\ 0 \end{bmatrix} e^{Ah} \int_0^h e^{(j\theta_k - A)\tau}\, d\tau B_1 \\
(\tilde{C})_k &= h^{-\frac{1}{2}} [C_1 \;\; D_{12}] \int_0^h \exp\left(\begin{bmatrix} A - j\theta_k & B_2 \\ 0 & -j\theta_k \end{bmatrix} \tau \right) d\tau \cdot \\
& \qquad \begin{bmatrix} I & 0 \\ D_{K_d} C_2 & C_{K_d} \end{bmatrix} \\
(\tilde{D})_{lk} &= h^{-1} C_1 \int_0^h e^{(A - j\theta_l)\tau} \int_0^\tau e^{(j\theta_k - A)\eta}\, d\eta\, d\tau B_1,
\end{aligned}
$$

where the sequence θ_k is defined above.

The matrix $\underline{M}_n(e^{j\omega_0})$ is the trucation of $M(e^{j\omega_0})$ and is formed by

$$
\underline{M}_n(e^{j\omega_0}) = \begin{bmatrix} (\tilde{C})_0 \\ \vdots \\ (\tilde{C})_n \end{bmatrix} e^{j\omega_0} (I - A_d e^{j\omega_0})^{-1} [(\tilde{B})_0 \; \ldots \; (\tilde{B})_n]
$$

$$+\begin{bmatrix} (\tilde{D})_{00} & \dots & (\tilde{D})_{0n} \\ \vdots & \ddots & \vdots \\ (\tilde{D})_{n0} & \dots & (\tilde{D})_{nn} \end{bmatrix}$$

where these matrix blocks are defined above.

Appendix B

Proof of Proposition 5.4

The LHS inequality is immediate from Theorem 5.1.

To show the RHS first set $n \geq 0$. From Corollary 4.16 there exists a sequence $\{\Delta_k\}_{k=0}^{\infty}$ in $\mathcal{X}$ with $\bar{\sigma}(\Delta_k) \leq 1$ so that

$$\mu_{\Delta}(M(e^{j\omega_0})) = \text{rad}\Big(\sum_{k=0}^{\infty}(\tilde{B})_k\Delta_k(\tilde{C})_k\, e^{j\omega_0}(I - e^{j\omega_0}A_d)^{-1}\Big).$$

For convenience set $T(e^{j\omega_0}) := e^{j\omega_0}(I - e^{j\omega_0}A_d)^{-1}$. By a direct application of (5.7) we have

$$\Big|\,\text{rad}\Big(\sum_{k=0}^{\infty}(\tilde{B})_k\Delta_k(\tilde{C})_k\, T(e^{j\omega_0})\Big) - \text{rad}\Big(\sum_{k=0}^{n}(\tilde{B})_k\Delta_k(\tilde{C})_k\, T(e^{j\omega_0})\Big)\,\Big| \leq b_n, \tag{B.1}$$

where

$$b_n := (2m_n')^{1-1/\tilde{n}}(\tilde{n}\bar{\sigma}(\sum_{k=n+1}^{\infty}(\tilde{B})_k\Delta_k(\tilde{C})_k\, T(e^{j\omega_0}))^{1/\tilde{n}}$$

and

$$m_n' = \max\{\bar{\sigma}\Big(\sum_{k=0}^{\infty}(\tilde{B})_k\Delta_k(\tilde{C})_k\, T(e^{j\omega_0})\Big),\ \bar{\sigma}\Big(\sum_{k=0}^{n}(\tilde{B})_k\Delta_k(\tilde{C})_k\, T(e^{j\omega_0})\Big)\}.$$

Now, for $\nu \geq 0$ the following inequality is satisfied

$$\bar{\sigma}\Big(\sum_{k=\nu}^{\infty}(\tilde{B})_k\Delta_k(\tilde{C})_k\, T(e^{j\omega_0})\Big) \leq \sum_{k=\nu}^{\infty}\bar{\sigma}((\tilde{B})_k)\bar{\sigma}((\tilde{C})_k\, T(e^{j\omega_0})) =: r_{\nu}.$$

Therefore, by the definition of b_n and m'_n we have that

$$b_n \leq (2r_{-1})^{1-1/\tilde{n}}(\tilde{n}r_{n+1})^{1/\tilde{n}} =: \epsilon_n.$$

From Proposition 5.3 we have that

$$\mu_{\underline{\Delta}_n}(\underline{M}_n) \geq \operatorname{rad}\left(\sum_{k=0}^{n} (\tilde{B})_k \Delta_k (\tilde{C})_k \, T(e^{j\omega_0}) \right)$$

and therefore by (B.1) we have

$$\mu_{\Delta}(M(e^{j\omega_0})) \leq \mu_{\underline{\Delta}_n}(\underline{M}_n(e^{j\omega_0})) + \epsilon_n \qquad \blacksquare$$

Appendix C

State space for $\bar{M}_n$

Here we provide formulae for the component matrices of $\overline{M}_n$. All have been derived using the method outlined in Subsection 4.3. They are conveniently expressed by the following matrices.

$$
\begin{aligned}
A_V &= \begin{bmatrix} A & B_1B_1^* & 0 \\ -C_1^*C_1 & -A^* & -C_1^*C_1 \\ 0 & 0 & A \end{bmatrix} \\
A_W &= \begin{bmatrix} A & B_1B_1^* & 0 & 0 \\ -C_1^*C_1 & -A^* & -C_1^*C_1 & -C_1^*D_{12} \\ 0 & 0 & A & B_2 \\ 0 & 0 & 0 & 0 \end{bmatrix} \\
A_X &= \begin{bmatrix} 0 & -D_{12}^*C_1 & -B_2^* & -D_{12}^*D_{12} \\ 0 & A & B_1B_1^* & B_2 \\ 0 & -C_1^*C_1 & -A^* & -C_1^*D_{12} \\ 0 & 0 & 0 & 0 \end{bmatrix} \\
A_Y &= \begin{bmatrix} A & B_1B_1^* \\ -C_1^*C_1 & -A^* \end{bmatrix} \\
A_Z &= \begin{bmatrix} A & B_1B_1^* & B_2 \\ -C_1^*C_1 & -A^* & -C_1^*D_{12} \\ 0 & 0 & 0 \end{bmatrix} \\
B_C &= h^{-\frac{1}{2}} \left\{ I - \exp(\begin{bmatrix} A & B_2 \\ 0 & 0 \end{bmatrix} h)\, e^{j\omega_0} \right\} \\
&\quad \cdot \begin{bmatrix} I & 0 & -e^{j\omega_0}(I - e^{j\omega_0}e^{Ah})^{-1} \\ D_{K_d}C_2 & C_K & 0 \end{bmatrix}
\end{aligned}
$$

$$C_B = h^{-\frac{1}{2}} \begin{bmatrix} I \\ 0 \\ I \end{bmatrix} (e^{-j\omega_0} - e^{Ah})$$

$$J_k = e^{j\omega_0} e^{Ah} \int_0^h e^{(j\theta_k - A)\tau}\, d\tau$$

The matrices $\hat{A}$, $\hat{B}_1$, $\hat{C}_1$, and $\hat{D}_{11}$ are given by their definition in (5.18). The remaining maps are given in block form below, with the number of block entries indicated in the right-hand column.

$$(\hat{D}_{12}T\hat{C}_2)_k = C_1 J_k [I \;\; 0 \;\; 0 \;\; 0] F_n(A_W) \begin{bmatrix} 0 \\ I \end{bmatrix} B_C \qquad (n+1)\times 1$$

$$(\hat{D}_{12}T\hat{D}_{21})_{kl} = C_1 J_k [I \;\; 0 \;\; 0] F_n(A_V) \begin{bmatrix} 0 \\ 0 \\ I \end{bmatrix} J_l B_1 \qquad (n+1)\times(n+1)$$

$$(\hat{B}_2 T\hat{D}_{21})_{kl} = C_B [I \;\; 0 \;\; 0] F_n(A_V) \begin{bmatrix} 0 \\ 0 \\ I \end{bmatrix} J_l B_1 \qquad 1\times(n+1)$$

$$\hat{B}_2 R \hat{B}_2^* = C_B [I \;\; 0] F_n(A_Y) \begin{bmatrix} 0 \\ -I \end{bmatrix} C_B^*$$

$$(\hat{B}_2 R \hat{D}_{12}^*)_l = C_B [I \;\; 0] F_n(A_Y) \begin{bmatrix} 0 \\ -I \end{bmatrix} J_l^* C_1^* \qquad 1\times(n+1)$$

$$(\hat{D}_{12} R \hat{D}_{12}^*)_{kl} = C_1 J_k [I \;\; 0] F_n(A_Y) \begin{bmatrix} 0 \\ -I \end{bmatrix} J_l^* C_1^* \qquad (n+1)\times(n+1)$$

$$\hat{C}_2^* N \hat{C}_2 = B_C^* \begin{bmatrix} 0 & 0 & I & 0 \\ I & 0 & 0 & 0 \end{bmatrix} F_n(A_X) \cdot \begin{bmatrix} 0 & I & 0 & 0 \\ 0 & 0 & 0 & I \end{bmatrix}^* B_C$$

$$(\hat{D}_{21}^* N \hat{C}_2)_k = B_1^* J_k^* [0 \;\; I \;\; 0] F_n(A_Z) \begin{bmatrix} I & 0 \\ 0 & 0 \\ 0 & I \end{bmatrix} B_C \qquad (n+1)\times 1$$

$$(\hat{D}_{21}^* N \hat{D}_{21})_{kl} = B_1^* J_k^* [0 \;\; I] F_n(A_Y) \begin{bmatrix} I \\ 0 \end{bmatrix} J_l B_1 \qquad (n+1)\times(n+1)$$

When the transfer function $C_1(Is - A)^{-1}B_1$ has no poles on the imaginary axis the above repesentation becomes much simpler. This is because the decomposition in Lemma 5.15 is valid for all k, l and ω_0. Equation

(5.18) can be rewritten as

$$M_n = \begin{bmatrix} \tilde{C}_1 & \check{C}_1 \\ \tilde{C}_2 & \check{C}_2 \end{bmatrix} e^{j\omega_0}(I - e^{j\omega_0} \begin{bmatrix} A_d & 0 \\ 0 & 0 \end{bmatrix})^{-1} \begin{bmatrix} \tilde{B}_1 & \tilde{B}_2 \\ \check{B}_1 & \check{B}_2 \end{bmatrix} + \begin{bmatrix} \hat{G}_{111} & 0 \\ 0 & \hat{G}_{112} \end{bmatrix},$$

and $\overline{M}_n$ in (5.22) becomes

$$\left[\begin{array}{c|cc} \hat{A} + \hat{B}_2 T \hat{C}_2 & \hat{B}_1 & \bar{B}_2 \\ \hline \hat{C}_1 & \hat{G}_{111} & 0 \\ \bar{C}_2 & 0 & 0 \end{array}\right],$$

where $\hat{G}_{111}$ =diag($(\hat{G}_{11})_0, \ldots, (\hat{G}_{11})_n$) defined in Lemma 5.15. Here $\bar{B}_2$ and $\bar{C}_2$ satisfy

$$\begin{aligned} \hat{B}_2 R \hat{B}_2^* &= \bar{B}_2 \bar{B}_2^* \\ \hat{C}_2^* N \hat{C}_2 &= \bar{C}_2^* \bar{C}_2, \end{aligned}$$

where the LHS matrices are provided above.

Appendix D

State Space for Section 6.2

Formulae are provided for computing the objects of Subsection 6.2.2. For simplicity we work with the scaling matrix $D_0 = I$ and $\lambda = 1$; to get formulae for general D_0 and λ set C_1 to $D_0^{\frac{1}{2}} C_1$, D_{12} to $D_0^{\frac{1}{2}} D_{12}$, and B_1 to $B_1(\lambda D_0)^{-\frac{1}{2}}$ in the derived formulae.

An additional note: in the example algorithm of Section 6.3 expressions were required for $U^* D_0^{\frac{1}{2}} \check{M}^* E_k \check{M} D_0^{-\frac{1}{2}} U$ and $U^* D_0^{-\frac{1}{2}} E_k D_0^{-\frac{1}{2}} U$. They can be obtained by

$$
\begin{aligned}
(U^* D_0^{-\frac{1}{2}}) \check{M}^* E_k \check{M} (D_0^{-\frac{1}{2}} U) &= d_k^{-1} U^* (D_0^{-\frac{1}{2}} \check{M}^* D_0^{\frac{1}{2}}) E_k (D_0^{\frac{1}{2}} \check{M} D_0^{-\frac{1}{2}}) U \\
(U^* D_0^{-\frac{1}{2}}) E_k (D_0^{-\frac{1}{2}}) U &= d_k^{-1} U^* E_k U
\end{aligned}
$$

since $D_0^{-\frac{1}{2}} E_k D_0^{-\frac{1}{2}} = d_k^{-1} E_k$, and using the state space expressions for the RHS derived below.

We remark here that many of the calculations below are part of the formulae associated with the $\mathcal{H}_\infty$-optimal sampled-data problem. See [7]. In the sequel we assume that (A, B_1, C_1) is a minimal realization.

<u>(i) Compute $\tilde{R}_1 \tilde{P}$:</u>

From (6.23) we have that

$$\tilde{R}_1 \tilde{P} := (I - \check{D}^* \check{D})^{-1} [\check{D}^* \check{C} \quad \check{B}^*] \tag{D.1}$$

where we have the *standing* assumption that $1 \notin \mathrm{spec}(\check{D}^*\check{D})$. The operator equation $y = \check{D}z$ can be realized by

$$\begin{aligned} \dot{x}_1(t) &= Ax_1(t) + B_1 z(t), \quad x_1(0) = 0 \\ y(t) &= C_1 x_1(t). \end{aligned}$$

Its adjoint $\check{D}^*$ is described by $(\check{D}^* y)(t) = -B_1^* \int_t^h e^{-A^*(t-\eta)} C_1^* y(\eta)\, d\eta = w(t)$, which has state space form

$$\begin{aligned} \dot{x}_2(t) &= -A^* x_2(t) - C_1^* y(t), \quad x_2(h) = 0 \\ w(t) &= B_1^* x_2(t). \end{aligned}$$

Hence, $w = (I - \check{D}^*\check{D})z$ has a realization in terms of the following boundary value problem:

$$\begin{aligned} \begin{bmatrix} \dot{x}_1(t) \\ \dot{x}_2(t) \end{bmatrix} &= \begin{bmatrix} A & 0 \\ -C_1^* C_1 & -A^* \end{bmatrix} \begin{bmatrix} x_1(t) \\ x_2(t) \end{bmatrix} + \begin{bmatrix} B_1 \\ 0 \end{bmatrix} z(t), \quad \begin{bmatrix} x_1(0) \\ x_2(h) \end{bmatrix} = 0 \\ w(t) &= [0 \quad -B_1^*] \begin{bmatrix} x_1(t) \\ x_2(t) \end{bmatrix} + z(t). \end{aligned}$$

To get a realization of $(I - \check{D}^*\check{D})^{-1}$ we solve for z above in terms of w to get

$$\begin{aligned} \begin{bmatrix} \dot{x}_1(t) \\ \dot{x}_2(t) \end{bmatrix} &= \begin{bmatrix} A & B_1 B_1^* \\ -C_1^* C_1 & -A^* \end{bmatrix} \begin{bmatrix} x_1(t) \\ x_2(t) \end{bmatrix} + \begin{bmatrix} B_1 \\ 0 \end{bmatrix} w(t), \quad \begin{bmatrix} x_1(0) \\ x_2(h) \end{bmatrix} = 0 \\ z(t) &= [0 \quad B_1^*] \begin{bmatrix} x_1(t) \\ x_2(t) \end{bmatrix} + w(t). \end{aligned} \tag{D.2}$$

This boundary value problem always has a solution under our assumption that $1 \notin \mathrm{spec}(\check{D}^*\check{D})$; see [75]. In the sequel it is convenient to have the following definitions.

$$\begin{aligned} H &:= \begin{bmatrix} A & B_1 B_1^* \\ -C_1^* C_1 & -A^* \end{bmatrix} & E(t) &:= e^{Ht} =: \begin{bmatrix} E_{11}(t) & E_{12}(t) \\ E_{21}(t) & E_{22}(t) \end{bmatrix} \\ L(t) &:= \int_0^t e^{H\eta}\, d\eta & Y(t) &:= \int_0^t e^{A\eta}\, d\eta. \end{aligned} \tag{D.3}$$

We remark from [75] that $1 \notin \operatorname{spec}(\check{D}^*\check{D})$ if and only if $E_{22}(h)$ is nonsingular.

To get $\tilde{R}_1\tilde{P}$ in (D.1) we first calculate $\tilde{R}_1\check{D}^*\check{C}$. From the realization of $\check{C}q$ in (A.2) we have

$$\begin{aligned} \dot{x}_0(t) &= Ax_0(t) + B_2u_0, \\ w(t) &= C_1x_0(t) + D_{12}u_0. \end{aligned} \qquad \begin{bmatrix} x_0(0) \\ u_0 \end{bmatrix} := \begin{bmatrix} I & 0 \\ D_{K_d}C_2 & C_{K_d} \end{bmatrix} \begin{bmatrix} q_1 \\ q_2 \end{bmatrix} \tag{D.4}$$

Also by an argument similar to that which obtained (D.2) the operator equation $z = \tilde{R}_1\check{D}^*w = (I - \check{D}^*\check{D})^{-1}\check{D}^*w$ has realization

$$\begin{bmatrix} \dot{x}_1(t) \\ \dot{x}_2(t) \end{bmatrix} = \begin{bmatrix} A & B_1B_1^* \\ -C_1^*C_1 & -A^* \end{bmatrix} \begin{bmatrix} x_1(t) \\ x_2(t) \end{bmatrix} + \begin{bmatrix} 0 \\ -C_1^* \end{bmatrix} w(t), \qquad \begin{bmatrix} x_1(0) \\ x_2(h) \end{bmatrix} = 0 \tag{D.5}$$

$$z(t) = [0 \quad B_1^*] \begin{bmatrix} x_1(t) \\ x_2(t) \end{bmatrix}. \tag{D.6}$$

Now, simply composing the above systems, and removing the uncontrollable and unobservable states, we get that $z = \tilde{R}_1\check{D}^*\check{C}q$ is given by

$$\begin{bmatrix} \dot{x}_1(t) \\ \dot{x}_2(t) \end{bmatrix} = \begin{bmatrix} A & B_1B_1^* \\ -C_1^*C_1 & -A^* \end{bmatrix} \begin{bmatrix} x_1(t) \\ x_2(t) \end{bmatrix} + \begin{bmatrix} B_2 \\ -C_1^*D_{12} \end{bmatrix} u_0,$$

$$\text{with boundary conditions} \begin{bmatrix} x_1(0) \\ x_2(h) \end{bmatrix} = \begin{bmatrix} q_1 \\ 0 \end{bmatrix},$$

$$z(t) = [0 \quad B_1^*] \begin{bmatrix} x_1(t) \\ x_2(t) \end{bmatrix}, \tag{D.7}$$

where $x_1(0)$ and u_0 are as defined in (D.4). To convert this boundary value problem to an initial value problem, we evaluate the system state at $t = h$, using the boundary condition, and solve for $x_2(0)$ to get

$$x_2(0) = -E_{22}^{-1}(h)E_{21}q_1 + [0 \quad -E_{22}^{-1}]L(h)Z_1u_0$$

where

$$Z_1 := \begin{bmatrix} B_2 \\ -C_1^*D_{12} \end{bmatrix}.$$

Hence, we can convert $z = \tilde{R}_1 \check{D}^* \check{C} q$ described in (D.7) to

$$z(t) = [0 \quad B_1^*] \left\{ E(t) Z_2 + L(t) Z_3 \right\} q \tag{D.8}$$

where

$$\begin{aligned} Z_2 &:= \begin{bmatrix} I & 0 \\ -E_{22}^{-1}(h) E_{21} & [0 \quad -E_{22}^{-1}] L(h) Z_1 \end{bmatrix} \begin{bmatrix} I & 0 \\ D_{K_d} C_2 & C_{K_d} \end{bmatrix} \\ Z_3 &:= [0 \quad Z_1] \begin{bmatrix} I & 0 \\ D_{K_d} C_2 & C_{K_d} \end{bmatrix}. \end{aligned}$$

Using the expression for $\check{B}^*$ in (A.2), and a similar state space argument to that above, it is routine to verify that $z = \tilde{R}_1 \check{B}^* \nu$ can be represented by

$$z(t) = [0 \quad B_1^*] E(t) Z_4 \nu, \qquad \text{where} \quad Z_4 := \begin{bmatrix} 0 \\ E_{22}^{-1}(h) \end{bmatrix} [I \quad 0].$$

Combining the last two equations

$$\tilde{R}_1 \tilde{P} = [\tilde{R}_1 \check{D}^* \check{C} \quad \tilde{R}_1 \check{B}^*] = [0 \quad B_1^*] \left\{ E(t) [Z_2 \quad Z_4] + L(t) [Z_3 \quad 0] \right\}. \tag{D.9}$$

<u>(ii) Compute $\check{M} \tilde{R}_1 \tilde{P}$</u>

From (3.5) we have that

$$\check{M} \tilde{R}_1 \tilde{P} = \left\{ \check{C} z (Iz - A_d)^{-1} \check{B} + \check{D} \right\} \tilde{R}_1 \tilde{P}.$$

Now consider first the term $\check{D} \tilde{R}_1 \tilde{P}$. Using the identities

$$\begin{aligned} \int_0^t e^{A(t-\eta)} [0 \quad B_1 B_1^*] E(\eta)\, d\eta &= [I \quad 0] E(t) - e^{At} [I \quad 0] \\ \int_0^h e^{A(t-\tau)} [0 \quad B_1 B_1^*] L(\tau)\, d\tau &= [I \quad 0] L(t) - Y(t) [I \quad 0], \end{aligned}$$

and the definition of $\check{D}$ in (A.2), it is a routine task to verify from (D.9) that

$$\begin{aligned} \check{D} \tilde{R}_1 \tilde{P} &= C_1 \left\{ [I \quad 0] E(t) - e^{At} [I \quad 0] \right\} [Z_2 \quad Z_4] \\ &\quad + C_1 \left\{ [I \quad 0] L(t) - Y(t) [I \quad 0] \right\} [Z_3 \quad 0] \\ &=: C_1 Z_5(t). \end{aligned} \tag{D.10}$$

Noting in (A.2) that $\check{B}$ and $\check{D}$ are defined by the same integral it is clear that

$$\check{B}\tilde{R}_1\tilde{P} = \begin{bmatrix} I \\ 0 \end{bmatrix} Z_5(h). \tag{D.11}$$

Using this and the integral form of $\check{C}$ given in (A.2) we have

$$\begin{aligned}
&\check{C}e^{j\omega}(I - e^{j\omega}A_d)^{-1}\check{B}\tilde{R}_1\tilde{P} \\
&= [C_1e^{At} \quad C_1Y(t)B_2 + D_{12}] \cdot \\
&\quad \begin{bmatrix} I & 0 \\ D_{K_d}C_2 & C_{K_d} \end{bmatrix} e^{j\omega}(I - e^{j\omega}A_d)^{-1} \begin{bmatrix} I \\ 0 \end{bmatrix} Z_5(h) \\
&=: [C_1e^{At} \quad C_1Y(t)B_2 + D_{12}] \begin{bmatrix} Z_6 \\ Z_7 \end{bmatrix}.
\end{aligned}$$

From this and (D.10) we have that

$$\begin{aligned}
\check{M}\tilde{R}_1\tilde{P} \;=\; & C_1[I \quad 0]\left\{E(t)[Z_2 \quad Z_4] + L(t)[Z_3 \quad 0]\right\} \\
& + C_1e^{At}\left\{Z_6 - [I \quad 0][Z_2 \quad Z_4]\right\} \\
& + C_1Y(t)\left\{B_2Z_6 - [I \quad 0][Z_3 \quad 0]\right\} + D_{12}Z_7.
\end{aligned} \tag{D.12}$$

<u>(iii) Compute $\tilde{P}^*\tilde{R}_1\tilde{P}$:</u>

By definition of $\tilde{P}$ we have that

$$\tilde{P}^*\tilde{R}_1\tilde{P} = \begin{bmatrix} \check{C}^*\check{D}\tilde{R}_1\check{D}^*\check{C} & \check{C}^*\check{D}\tilde{R}_1\check{B}^* \\ \check{B}\tilde{R}_1\check{D}^*\check{C} & \check{B}\tilde{R}_1\check{B}^* \end{bmatrix}.$$

From (D.11) we have $\check{B}\tilde{R}_1\tilde{P} = [\check{B}\tilde{R}_1\check{D}^*\check{C} \quad \check{B}\tilde{R}_1\check{B}^*]$, so it remains to calculate $\check{C}^*\check{D}\tilde{R}_1\check{D}^*\check{C}$: we will calculate it from

$$\check{C}^*\check{D}\tilde{R}_1\check{D}^*\check{C} = \check{C}^*(I - \check{D}\check{D}^*)^{-1}\check{C} - \check{C}^*\check{C}.$$

For the moment we concentrate on evaluating $\check{C}^*(I - \check{D}\check{D}^*)^{-1}\check{C}$. By virtually the same argument as (D.4) to (D.7) we get that $z(t) = (I - \check{D}\check{D}^*)^{-1}\check{C}q$

obeys

$$\begin{bmatrix} \dot{x}_1(t) \\ \dot{x}_2(t) \end{bmatrix} = \begin{bmatrix} A & B_1B_1^* \\ -C_1^*C_1 & -A^* \end{bmatrix} \begin{bmatrix} x_1(t) \\ x_2(t) \end{bmatrix} + \begin{bmatrix} B_2 \\ -C_1^*D_{12} \end{bmatrix} u_0,$$

$$\text{with boundary conditions } \begin{bmatrix} x_1(0) \\ x_2(h) \end{bmatrix} = \begin{bmatrix} q_1 \\ 0 \end{bmatrix},$$

$$z(t) = [C_1 \;\; 0] \begin{bmatrix} x_1(t) \\ x_2(t) \end{bmatrix}, \tag{D.13}$$

where $u_0 = [D_{K_d}C_2 \;\; C_{K_d}]q$ and $x_1(0) = [I \;\; 0]q$. This system has the same state equation as (D.7) and therefore using the same steps as led to (D.8) we have

$$(I - \check{D}\check{D}^*)^{-1}\check{C} = [C_1 \;\; 0]\left\{E(t)Z_2 + L(t)Z_3\right\}.$$

To complete the evaluation we use the following general identity which holds for any square matrices F and X, and appropriately dimensioned matrix Q:

$$\int_0^t e^{F\eta}Qe^{X(t-\eta)}\,d\eta = [I \;\; 0]\exp(\begin{bmatrix} F & Q \\ 0 & X \end{bmatrix} t) \begin{bmatrix} 0 \\ I \end{bmatrix}. \tag{D.14}$$

This is straightforward to verify and [67] considers computational issues associated with this formula. Using this identity we can write

$$(I - \check{D}\check{D}^*)^{-1}\check{C} = [C_1 \;\; 0]\,[I \;\; 0]\exp\left(\begin{bmatrix} H & I \\ 0 & 0 \end{bmatrix} t\right) \begin{bmatrix} Z_2 \\ Z_3 \end{bmatrix}$$

Also, based on (A.2) and the above identity

$$\check{C} = [C_1 \;\; D_{12}]\exp\left(\begin{bmatrix} A & B_2 \\ 0 & 0 \end{bmatrix} t\right) \begin{bmatrix} I & 0 \\ D_{K_d}C_2 & C_{K_d} \end{bmatrix}.$$

Hence, we can use (D.14) to evaluate $\check{C}^*\check{D}\tilde{R}_1\check{D}^*\check{C} = \check{C}^*(I - \check{D}\check{D}^*)^{-1}\check{C} - \check{C}^*\check{C}$ from the last two expressions.

(iv) Evaluate $\tilde{P}^*\tilde{R}_1E_k\tilde{R}_1\tilde{P}^*$ and $\tilde{P}^*\tilde{R}_1\check{M}^*E_k\check{M}\tilde{R}_1\tilde{P}^*$

Here again the approach is to employ (D.14); we use it to get from (D.9) to

$$\tilde{R}_1\tilde{P} = [0 \;\; B_1^*]\,[I \;\; 0]\exp\left(\begin{bmatrix} H & I \\ 0 & 0 \end{bmatrix} t\right) \begin{bmatrix} Z_2 & Z_4 \\ Z_3 & 0 \end{bmatrix}.$$

Then $\tilde{P}^*\tilde{R}_1E_k\tilde{R}_1\tilde{P}$ and $\tilde{P}^*\tilde{R}_1^2\tilde{P}$ can be evaluated using the identity. Also, note that

$$\tilde{P}^*\tilde{R}_1^2\tilde{P} = \sum_{k=1}^{d+1} \tilde{P}^*\tilde{R}_1E_k\tilde{R}_1\tilde{P}.$$

Similarly, from (D.12) and (D.14) we have

$$\begin{aligned}
\check{M}\tilde{R}_1\tilde{P} &= [C_1 \;\; 0]\,[I \;\; 0]\exp\left(\begin{bmatrix} H & I \\ 0 & 0 \end{bmatrix} t\right)\begin{bmatrix} Z_2 & Z_4 \\ Z_3 & 0 \end{bmatrix} \\
&\quad + [C_1 \;\; D_{12}Z_7]\exp\left(\begin{bmatrix} A & Z_8 \\ 0 & 0 \end{bmatrix} t\right)\begin{bmatrix} I & 0 \\ D_{K_d}C_2 & C_{K_d} \end{bmatrix}\begin{bmatrix} Z_9 \\ I \end{bmatrix},
\end{aligned}$$

where $Z_8 := B_2Z_7 - [I \;\; 0][Z_3 \;\; 0]$ and $Z_7 = Q_1 - [I \;\; 0][Z_2 \;\; Z_4]$. Then $\tilde{P}^*\tilde{R}_1\check{M}^*E_k\check{M}\tilde{R}_1\tilde{P}^*$ can be computed using the integral identity (D.14).

Appendix E

Proof of Lemma 6.10

To prove Lemma 6.10, we require an interpolation result which is a generalization of Lemma 6.1 from one interpolation point to a finite number of them:

Corollary E.1 *Given n distinct frequencies $\omega_1 < \ldots < \omega_n$ that satisfy hypothesis (i) of Lemma 6.10, and a set of operators $Q_1, \ldots, Q_n$ in $\mathcal{UL}(\mathcal{K}_2)$. There exists a causal operator $\mathbf{Q}$ in $\mathcal{UL}_{\mathcal{A}}$ so that its transfer function satisfies*

$$\check{Q}(e^{j\omega_l}) = Q_l, \qquad \text{for } 1 \leq l \leq n.$$

The following proof closely parallels the proof of Lemma 6.1.

Proof By a discrete version of Lemma 4.9 there exist n contractive functions f_k in $\mathcal{A}$ so that

$$f_k(e^{j\omega_l}) = \begin{cases} 1, & k = l \\ 0, & 1 \leq l \leq n,\ l \neq k. \end{cases},$$

and satisfy $f_k(z^*) = f_k(z)^*$. Now set,

$$\check{Q}(z) = \sum_{k=1}^{n} z f_k(z) \cdot$$

$$\left\{ \left(\frac{e^{j\omega_k} - (1 + 1/\alpha)e^{-j\omega_k}}{e^{j2\omega_k} - 1} \frac{z - e^{-j\omega_k}}{z - (1 + 1/\alpha)e^{-j\omega_k}} \frac{Q_k}{1 + \alpha(1 - e^{-j\omega_k} z)} \right) + \left(\frac{e^{-j\omega_k} - (1 + 1/\alpha)e^{j\omega_k}}{e^{-j2\omega_k} - 1} \frac{z - e^{j\omega_k}}{z - (1 + 1/\alpha)e^{j\omega_k}} \frac{(Q_k^T)^*}{1 + \alpha(1 - e^{j\omega_k} z)} \right) \right\}$$

where $\alpha > 0$ is chosen sufficiently large so that $\|\check{Q}\|_\infty < 1$. The corresponding operator $\mathbf{Q} \in \mathcal{UL}_{\mathcal{A}}$, and is causal by the identical argument used in the proof of Lemma 6.1. ■

Proof of Lemma 6.10

Step 1: Construct $\mathbf{\Delta}$ that satisfies the limit

We will construct a perturbation $\tilde{\Delta} \in \mathfrak{L}(\ell_2)$ that corresponds to $\mathbf{\Delta}$. First, if the sequence b^l is identically zero then we simply set $\tilde{\Delta} = 0$. Otherwise, by assumption $\sum_{l=1}^n \|b^l\|_{\mathcal{K}_2}^2 < \sum_{l=1}^n \|a^l\|_{\mathcal{K}_2}^2$ and we define

$$\eta^2 := \frac{\sum_{l=1}^n \|a^l\|_{\mathcal{K}_2}^2}{\sum_{l=1}^n \|b^l\|_{\mathcal{K}_2}^2}, \tag{E.1}$$

and set ϵ to some value satisfying $0 < \epsilon < \sqrt{\eta} - 1$.

By Corollary E.1 there exists a causal operator $\tilde{F}$ mapping $\ell_2^m \to (\overset{n}{\underset{l=1}{\oplus}} \ell_2^m)$ which has the form

$$\tilde{F} =: \begin{bmatrix} \tilde{F}_1 \\ \vdots \\ \tilde{F}_n \end{bmatrix},$$

so that $\|\tilde{F}\|_{\ell_2 \to \ell_2} < 1 + \epsilon$ and the associated transfer functions $\check{F}_k$ satisfy

$$\check{F}_k(e^{j\omega_l}) = \begin{cases} I, & l = k \\ 0, & 1 \le l \le n,\ l \ne k \end{cases}$$

for $1 \le k \le n$.

Also, define the time varying operator $\tilde{Y} = \text{diag}(\tilde{Y}_1, \ldots, \tilde{Y}_n)$ on $(\overset{n}{\underset{l=1}{\oplus}} \ell_2^m)$, that is defined for each $\tilde{u} \in \ell_2$ and $k \ge 0$ by

$$(\tilde{Y}_l \tilde{u})[k] := e^{j(\omega_1 - \omega_l)k} \tilde{u}[k],$$

for $1 \le l \le n$. Clearly, $\tilde{Y}$ is unitary and its adjoint is defined by $(\tilde{Y}_l^* \tilde{u})[k] := e^{j(\omega_l - \omega_n)k} \tilde{u}[k]$, the adjoint of each of its components.

Now set $a := (a_1, \dots, a_n) \in (\overset{n}{\underset{l=1}{\oplus}} \mathcal{K}_2^m)$ and $\tilde{a}^q := a\phi_{\omega_1}^q$. From the above definitions and Lemma 6.9 it is routine to demonstrate that

$$\frac{\|\tilde{a}^q - \tilde{Y}\tilde{F}\tilde{z}^q\|_{\ell_2}}{\|\tilde{z}^q\|_{\ell_2}} \xrightarrow{q\to\infty} 0. \tag{E.2}$$

Next, define the $b := (b_1, \dots, b_n) \in (\overset{n}{\underset{l=1}{\oplus}} \mathcal{K}_2^m)$ and the memoryless operator $\tilde{V}$ on $(\overset{n}{\underset{l=1}{\oplus}} \ell_2^m)$ by

$$(\tilde{V}\tilde{v})[k] := \frac{\langle a, \tilde{v}[k]\rangle_{\mathcal{K}_2}}{\|a\|_{\mathcal{K}_2}^2} b,$$

for $\tilde{v} \in \ell_2$. Clearly, $\|\tilde{V}\| = \eta^{-1}$ from (E.1). So, with $\tilde{b}^q := b\phi_{\omega_1}^q$, we have by (E.2) that

$$\frac{\|\tilde{b}^q - \tilde{V}\tilde{Y}\tilde{F}\tilde{z}^q\|_{\ell_2}}{\|\tilde{z}^q\|_{\ell_2}} \xrightarrow{q\to\infty} 0. \tag{E.3}$$

We require one more operator: by Corollary E.1 there exists an operator $\tilde{E} = [\tilde{E}_1 \dots \tilde{E}_n]$ mapping $(\overset{n}{\underset{l=1}{\oplus}} \ell_2^m)$ to ℓ_2^m, with $\|\tilde{E}\|_{\ell_2\to\ell_2} < 1 + \epsilon$, so that the transfer functions $\check{E}_k$ satisfy

$$\check{E}_k(e^{j\omega_l}) = \begin{cases} e^{-j\omega_l} I, & l = k \\ 0, & 1 \le l \le n,\ l \ne k \end{cases}$$

for $1 \le k \le n$.

From the definition of $\tilde{Y}^*$ and (E.3), it is straightforward to verify that

$$\frac{\|\tilde{w}^q - \tilde{U}\tilde{E}\tilde{Y}^*\tilde{V}\tilde{Y}\tilde{F}\tilde{z}^q\|_{\ell_2}}{\|\tilde{z}^q\|_{\ell_2}} \xrightarrow{q\to\infty} 0,$$

where $\tilde{U}$ is the unilateral shift on ℓ_2.

We set $\tilde{\Delta} = \tilde{U}\tilde{E}\tilde{Y}^*\tilde{V}\tilde{Y}\tilde{F}$ which clearly satisfies the limit of the claim. Furthermore, from the submultiplicative inequality and the fact that $\tilde{Y}$ and $\tilde{U}$ are contractive we have that

$$\|\tilde{\Delta}\| \le \|\tilde{E}\|\, \|\tilde{V}\|\, \|\tilde{F}\| < (1+\epsilon)^2\eta^{-1} < 1 \tag{E.4}$$

by (E.1) and the construction of the operators. With the operator $\tilde{\Delta}$ above on ℓ_2 we associate the map Δ on $\mathcal{L}_2$. This operator Δ is causal on $\mathcal{L}_2$ using the same argument as given in the proof of Lemma 6.1; because $\tilde{E}\tilde{Y}^*\tilde{V}\tilde{Y}\tilde{F}$ is causal on ℓ_2 and $\tilde{\Delta}$ is a shifted version of it.

Step 2: Bound the time variation of Δ

We now show that $\Delta \in \mathfrak{L}_{QP}(\nu)$ where $\nu = 2\sin(\frac{\omega_n - \omega_1}{2})$; namely that $\|\tilde{U}\tilde{\Delta} - \tilde{\Delta}\tilde{U}\| < \nu$. First observe that since $\tilde{E}$ and $\tilde{F}$ are shift invariant that $\tilde{U}\tilde{\Delta} - \tilde{\Delta}\tilde{U} = \tilde{E}\tilde{U}(\tilde{U}\tilde{Y}^*\tilde{V}\tilde{Y} - \tilde{Y}^*\tilde{V}\tilde{Y}\tilde{U})\tilde{F}$. By the submultiplicative inequality we get

$$\|\tilde{U}\tilde{\Delta} - \tilde{\Delta}\tilde{U}\| \le \|\tilde{E}\| \, \|\tilde{U}\tilde{Y}^*\tilde{V}\tilde{Y} - \tilde{Y}^*\tilde{V}\tilde{Y}\tilde{U}\| \, \|\tilde{F}\|. \tag{E.5}$$

Also, using the submultiplicative inequality and the fact that $\tilde{Y}$ is unitary we get that

$$\begin{aligned}
\|\tilde{U}\tilde{Y}^*\tilde{V}\tilde{Y} - \tilde{Y}^*\tilde{V}\tilde{Y}\tilde{U}\| &\le \|\tilde{Y}\tilde{U}\tilde{Y}^*\tilde{V} - \tilde{V}\tilde{Y}\tilde{U}\tilde{Y}^*\| \\
&= \|\tilde{Y}\tilde{U}\tilde{Y}^*\tilde{V} - \alpha\tilde{U}\tilde{V} + \alpha\tilde{U}\tilde{V} - \tilde{V}\tilde{Y}\tilde{U}\tilde{Y}^*\| \\
&= \|\tilde{Y}\tilde{U}\tilde{Y}^*\tilde{V} - \alpha\tilde{U}\tilde{V} + \alpha\tilde{V}\tilde{U} - \tilde{V}\tilde{Y}\tilde{U}\tilde{Y}^*\| \\
&\le 2\|\tilde{Y}\tilde{U}\tilde{Y}^* - \alpha\tilde{U}\| \, \|\tilde{V}\|,
\end{aligned}$$

for any scalar α. Now, $\tilde{Y}\tilde{U}\tilde{Y}^* = \operatorname{diag}(I, e^{j(\omega_1-\omega_2)}I, \ldots, e^{j(\omega_1-\omega_n)}I)\tilde{U}$. Therefore, setting $\alpha = \frac{1}{2}(1 - e^{j(\omega_1-\omega_n)})$ we have $\|\tilde{Y}\tilde{U}\tilde{Y}^* - \alpha\tilde{U}\| = \sin(\frac{\omega_n-\omega_1}{2})$. By this and (E.5) we have

$$\|\tilde{U}\tilde{\Delta} - \tilde{\Delta}\tilde{U}\| < \nu\|\tilde{E}\| \, \|\tilde{V}\| \, \|\tilde{F}\|$$

which is therefore bounded by ν from (E.4). ■

Remark E.2 *Note that in the above proof we have not shown that the perturbation Δ is real, and in fact, as constructed, it is not. A real Δ can be constructed, using a similar argument, augmenting the perturbation with appropriate adjoint operators. See [50] for the main idea in such an extension.*

Appendix F

The Hilbert-Schmidt Norm of $E_k\check{M}E_l$

The integral kernel $M(\tau, t)$ in (6.37) is given in terms of the following matrices:

$$\begin{aligned} A_{\tilde{C}} &:= \begin{bmatrix} A & B_2 \\ 0 & 0 \end{bmatrix} \\ B_{\tilde{C}} &:= \begin{bmatrix} I & 0 \\ D_{K_d}C_2 & C_{K_d} \end{bmatrix} \\ C_{\tilde{C}} &:= [C_1 \;\; D_{12}] \\ C_{\tilde{B}} &:= \begin{bmatrix} I \\ 0 \end{bmatrix}. \end{aligned}$$

These are easily verified from the state space form of $\check{M}$ given in Appendix A.

By the kernel form of the Hilbert-Schmidt norm in (6.36) we have that $\|E_k\check{M}E_l\|_{\text{HS}}$, for k and l in $\{1, \ldots, d+1\}$, is given by

$$\|E_k\check{M}E_l\|_{\text{HS}} = \int_0^h \int_0^h \text{tr}(\, E_l M^*(t, \tau) E_k^2 M(t, \tau) E_l \,)\, d\tau\, dt.$$

Now, define the matices

$$\begin{aligned} T &:= B_{\tilde{C}} e^{j\omega_0} (I - e^{j\omega_0} A_d)^{-1} C_{\tilde{B}} \\ B^l &:= B_1 E_l \end{aligned}$$

$$\begin{aligned} C^k &:= E_k C_1 \\ \underline{C}^k &:= E_k C_{\check{C}}. \end{aligned}$$

Using these definitions and the integral kernel in (6.37) we get $\|E_k\check{M}E_l\|_{\mathrm{HS}}$ by standard manipulations:

$$\begin{aligned} &\|E_k\check{M}E_l\|_{\mathrm{HS}}{}^2 \\ &= \operatorname{tr}\left\{T^* \int_0^h e^{A_{\check{C}}^* t}(\underline{C}^k)^* \underline{C}^k e^{A_{\check{C}} t}\, dt\, T \int_0^h e^{A\tau} B^k (B^k)^* e^{A^*\tau}\, d\tau\right\} \\ &+\operatorname{tr}\left\{T^* \int_0^h \int_0^t e^{A^*\tau}(C^k)^* C^k e^{A\tau}\, d\tau\, dt\, B^k (B^k)^*\right\} \\ &+\operatorname{Re}\left(2\operatorname{tr}\left\{T^* \int_0^h e^{A_{\check{C}}^* t}(\underline{C}^k)^* C^k e^{At} \int_0^t e^{-A\tau} B^k (B^k)^* e^{-A^*\tau}\, d\tau\, dt\, e^{A^* h}\right\}\right). \end{aligned}$$

These integrals can be evaluated explicitly, for each k and l, using the identity (D.14).

Appendix G

The S-Procedure

The main purpose here is to prove Proposition 6.8. In fact we will prove the more general result stated in Corollary G.5; see also Remark G.6. The main technical result of this section, stated in Theorem G.3, is that the so-called S-procedure is lossless for a finite number of LTI quadratic forms; this result was first proved for ℓ_2 by Megretski and Treil in [41], and the version of their proof adapted here is due to Paganini and Doyle [48].

To start we define the notion of a time-invariant quadratic form on ℓ_2: a mapping $\psi : \ell_2 \to \mathbb{R}$ is a *time-invariant quadratic form* if there exist two time-invariant operators $\tilde{X}$ and $\tilde{Y}$ in $\mathcal{L}(\ell_2)$ satisfying

$$\psi(\tilde{u}) = \|\tilde{X}\tilde{u}\|_2^2 - \|\tilde{Y}\tilde{u}\|_2^2$$

for each $\tilde{u} \in \ell_2$.

Recalling the definitions of time-invariance and the unilateral shift operator $\tilde{U}$ on ℓ_2, we state the following elementary lemma without proof.

Lemma G.1 *Suppose $\tilde{u}$, $\tilde{v} \in \ell_2$ and ψ is a time-invariant quadratic form on ℓ_2. Then*

$$\begin{aligned} &\text{(i)} \quad \psi(\tilde{U}\tilde{u}) = \psi(\tilde{u}) \\ &\text{(ii)} \quad \lim_{l\to\infty} \psi(\tilde{U}^l\tilde{u} + \tilde{v}) = \psi(\tilde{u}) + \psi(\tilde{v}). \end{aligned}$$

where $\tilde{U}$ is the unilateral shift operator on ℓ_2.

Given a collection $\psi_1, \ldots, \psi_n$ of time-invariant quadratic forms, we define the set

$$\nabla := \{(\psi_1(\tilde{u}), \ldots, \psi_n(\tilde{u})) \in \mathbb{R}^n : \tilde{u} \in \boldsymbol{\ell}_2, \|\tilde{u}\|_2 = 1\}. \tag{G.1}$$

Pertaining to this set we have a key technical lemma.

Lemma G.2 *Suppose $\psi_1, \ldots, \psi_n$ is a collection of time-invariant quadratic forms and ∇ is the corresponding subset of $\mathbb{R}^n$ defined by (G.1). Then the closure $\bar{\nabla}$ is both convex and compact.*

Proof First, $\bar{\nabla}$ is clearly compact since it is bounded.

To prove convexity we show that $\bar{\nabla}$ is equal to its convex hull; it is sufficient to demonstrate that $\bar{\nabla}$ contains the convex hull of ∇. Choose $p, q \in \nabla$ and $\lambda \in [0, 1]$. Then by definition there exist $\tilde{u}, \tilde{v} \in \boldsymbol{\ell}_2$, $\|\tilde{u}\|_2 = \|\tilde{v}\|_2 = 1$, satisfying

$$p = (\psi_1(\tilde{u}), \ldots, \psi_n(\tilde{u})), \quad q = (\psi_1(\tilde{v}), \ldots, \psi_n(\tilde{v})).$$

For convenience set $\tilde{w}^l = \sqrt{\lambda}\tilde{u} + \sqrt{1-\lambda}\tilde{U}^l\tilde{v}$. By Lemma G.1

$$\lim_{l \to \infty} \psi_k(\tilde{w}^l) = \lambda\psi_k(\tilde{u}) + (1-\lambda)\psi_k(\tilde{v}),$$

for each $1 \le k \le n$. Hence, $\lim_{l\to\infty}(\psi_1(\tilde{w}^l), \ldots, \psi_n(\tilde{w}^l)) = \lambda p + (1-\lambda)q$. Since $\|\tilde{w}^l\|_2 \to 1$ as $l \to \infty$ the LHS limit must be in $\bar{\nabla}$, and therefore $\bar{\nabla}$ contains the convex hull of ∇. ∎

Denote the positive quadrant by

$$\Pi^+ := \{x \in \mathbb{R}^n : x = (x_1, \ldots, x_n), x_k \ge 0\}.$$

The major result of the section can now be proved. The process of converting condition (i) below on quadratic forms to that of (ii) has come to be know as the S-procedure; see [41] for a historical account of this terminology. The following result shows that for LTI quadratic forms (i) and (ii) are equivalent; namely the S-procedure is so-called lossless. This result is not true in general for quadratic forms that are *not* LTI.

Theorem G.3 *Suppose* $\psi_1, \ldots, \psi_n$ *is a collection of time-invariant quadratic forms and* ∇ *is the corresponding subset of* $\mathbb{R}^n$ *defined by (G.1). Then the following statements are equivalent:*

(i) The inequality $\inf_{x \in \nabla, y \in \Pi^+} |x - y| > 0$ *is satisfied*

(ii) There exist real scalars $d_k > 0$ *such that*

$$d_1\psi_1(\tilde{u}) + \ldots + d_n\psi_n(\tilde{u}) \leq -d_{n+1}$$

for all $\tilde{u} \in \ell_2$ *with* $\|\tilde{u}\|_2 = 1$.

Proof That (ii) implies (i) is straightforward: suppose (i) does not hold; then the LHS of (ii) can be made as close to zero as desired, thus violating inequality (ii).

To prove (i) implies (ii) we first observe that (i) says that $\bar{\nabla}$ and Π^+ are strictly separated. The set Π^+ is convex and by Lemma G.2 so is $\bar{\nabla}$. Therefore, (see e.g. [54]) there is a hyperplane that strictly separates them: there exist real numbers d'_k such that

$$\begin{aligned} d'_1 y_1 + \cdots + d'_n y_n &> -d'_{n+1} \quad \text{for all } y \in \Pi^+ \\ d'_1 x_1 + \cdots + d'_n x_n &\leq -d'_{n+1} \quad \text{for all } x \in \bar{\nabla}. \end{aligned}$$

The zero element is in Π^+, and therefore by the first inequality $d'_{n+1} > 0$. Also, the standard basis for $\mathbb{R}^n$ is contained in Π^+ so the first inequality implies $d'_k \geq 0$, for $1 \leq k \leq n$, by scaling the basis.

Finally, recalling (Lemma G.2) that $\bar{\nabla}$ is compact and using the second inequality above, we can choose $d_k > 0$ sufficiently close to $d'_k \geq 0$ so that (ii) is satisfied. ■

With this result we apply it to a specific set of quadratic forms. Given a time invariant operator $\tilde{T} \in \mathfrak{L}(\ell_2)$, and the orthogonal projections E_k defined in (6.26), we define the time-invariant quadratic forms

$$\psi_k(\tilde{u}) = \|E_k\tilde{T}\tilde{u}\|_2^2 - \|E_k\tilde{u}\|_2^2 \tag{G.2}$$

for each $1 \leq k \leq n$. We can now state and prove a corollary to Theorem G.3. The scaling set $\tilde{\mathcal{D}}_{\Delta_{rp}}$ referred to is defined in (6.9).

Corollary G.4 *Given a time-invariant operator $\tilde{T}$ and the corresponding set of time-invariant quadratic forms defined in (G.2). Then* $\inf_{x\in\nabla,y\in\Pi^+}|x-y|>0$ *holds if and only if*

$$\inf_{D\in\tilde{\mathcal{D}}_{\Delta rp}} \|D\tilde{T}D^{-1}\|_{\ell_2\to\ell_2} < 1. \tag{G.3}$$

Proof By Theorem G.3 we have $\inf_{x\in\nabla,y\in\Pi^+}|x-y|>0$ is equivalent to the existence of scalars $d_k>0$ satisfying

$$d_1\psi_1(\tilde{u})+\ldots+d_n\psi_n(\tilde{u}) \le -d_{n+1} \tag{G.4}$$

for all $\tilde{u}\in\ell_2$ with $\|\tilde{u}\|_2=1$. Hence, we must show that (G.4) is equivalent to (G.3).

(Prove (G.4) implies (G.3)): for given $d_k>0$ set $D=\sum_{k=1}^n\sqrt{d_k}E_k\in\tilde{\mathcal{D}}_{\Delta_{rp}}$ and observe (G.4) is

$$\begin{aligned} -d_{n+1} &\ge \sum_{k=1}^n d_k(\|E_k\tilde{T}\tilde{u}\|_2^2-\|E_k\tilde{u}\|_2^2) \\ &= \|D\tilde{T}\tilde{u}\|_2^2-\|D\tilde{u}\|_2^2 \end{aligned} \tag{G.5}$$

since the operators E_k are orthogonal projections. Thus, for each $\tilde{u}\in\ell_2$ we have

$$\|D\tilde{T}D^{-1}\tilde{u}\|_2^2-\|\tilde{u}\|_2^2 \le -d_{n+1}\|D^{-1}\tilde{u}\|_2^2 \le -d_{n+1}\|D\|^{-2}\|\tilde{u}\|_2^2$$

which implies (G.3).

(Prove (G.3) implies (G.4)): This follows by reversing the above argument setting $d_k=|d_k'|$ where $D=\sum_{k=1}^n\sqrt{d_k}E_k\in\tilde{\mathcal{D}}_{\Delta_{rp}}$. ∎

Corollary G.5 *Suppose $\tilde{Q}$ is a time-invariant operator on ℓ_2. The inequality* $\inf_{D\in\tilde{\mathcal{D}}_{\Delta_{rp}}}\|D\tilde{Q}D^{-1}\|_{\ell_2\to\ell_2}>1$ *holds if and only if there exist nonzero $\tilde{u}\in\ell_2$ and $\gamma>1$ such that*

$$\gamma\|E_k\tilde{u}\|_2 \le \|E_k\tilde{Q}\tilde{u}\|_2 \quad \textit{for each } 1\le k\le n.$$

Proof (If): Suppose $\gamma > 1$ and nonzero $\tilde{u} \in \ell_2$ exist such that $\gamma^{-2}\|E_k\tilde{Q}\tilde{u}\|_2^2 - \|E_k\tilde{u}\|_2^2 \geq 0$ holds for each $1 \leq k \leq n$. With $\tilde{T} := \gamma^{-1}\tilde{Q}$ and $\bar{\nabla}$ defined as in (G.1) we have $\inf_{x\in\nabla, y\in\Pi^+} |x - y| = 0$. Invoking Corollary G.4 we conclude

$$1 \leq \inf_{D\in\tilde{\mathcal{D}}_{\Delta rp}} \|D\tilde{T}D^{-1}\|_{\ell_2\to\ell_2} = \gamma^{-1} \inf_{D\in\tilde{\mathcal{D}}_{\Delta rp}} \|D\tilde{Q}D^{-1}\|_{\ell_2\to\ell_2}.$$

(Only if): By hypothesis there exists $\beta > 1$ satisfying

$$\beta^{-1} \inf_{D\in\tilde{\mathcal{D}}_{\Delta rp}} \|D\tilde{Q}D^{-1}\|_{\ell_2\to\ell_2} = 1.$$

From Corollary G.4, setting $\tilde{T} := \beta^{-1}\tilde{Q}$, we have $\inf_{x\in\nabla, y\in\Pi^+} |x - y| = 0$. Therefore, for every $\epsilon > 0$ there exists $\tilde{u} \in \ell_2$, with $\|\tilde{u}\|_2 = 1$, such that

$$\beta^{-2}\|E_k\tilde{Q}\tilde{u}\|_2^2 - \|E_k\tilde{u}\|_2^2 \geq -\epsilon,$$

for each $1 \leq k \leq n$.

From the last condition there exists an infinite sequence $\tilde{u}^l \in \ell_2$, $\|\tilde{u}\|_2 = 1$, so that

$$\begin{aligned} 0 &\leq \lim_{l\to\infty}\{\beta^{-2}\|E_k\tilde{Q}\tilde{u}^l\|_2^2 - \|E_k\tilde{u}^l\|_2^2\} \\ &= \lim_{l\to\infty}\beta^{-2}\|E_k\tilde{Q}\tilde{u}^l\|_2^2 - \lim_{l\to\infty}\|E_k\tilde{u}^l\|_2^2, \end{aligned} \tag{G.6}$$

for each $1 \leq k \leq n$. Furthermore, without loss of generality we may assume $\tilde{u}^l$ satisfies the following condition: if $\lim_{l\to\infty}\|E_{k_0}\tilde{u}^l\|_2^2 = 0$ for any $1 \leq k_0 \leq n$, then

$$E_{k_0}\tilde{u}^l = 0 \quad \text{for each } l \geq 0. \tag{G.7}$$

To complete the proof, choose $1 < \gamma < \beta$. It is now routine to demonstrate using (G.6) and the condition in (G.7), that there exists an integer l_0 sufficiently large so that $\gamma^{-2}\|E_k\tilde{Q}\tilde{u}^{l_0}\|_2^2 - \|E_k\tilde{u}^{l_0}\|_2^2 \geq 0$ for each $1 \leq k \leq n$. Now set $\tilde{u} = \tilde{u}^{l_0}$. ■

The following comment connects the time domain conditions derived so far with their equivalent frequency domain conditions, which are more amenable to computation:

Remark G.6 *The time domain quantity* $\inf_{D\in\tilde{\mathcal{D}}_{\Delta rp}} \|D\tilde{Q}D^{-1}\|_{\ell_2\to\ell_2}$, *in Corollary G.4 and Corollary G.5, is equal to the frequency domain quantity* $\inf_{D\in\tilde{\mathcal{D}}_{\Delta rp}} \|D\check{Q}D^{-1}\|_{\infty}$. *In particular if* $\tilde{Q}$ *is a memoryless multiplication operator defined from* $Q \in \mathfrak{L}(\mathcal{K}_2)$, *then* $\inf_{D\in\tilde{\mathcal{D}}_{\Delta rp}} \|D\check{Q}D^{-1}\|_{\infty} = \inf_{D\in\tilde{\mathcal{D}}_{\Delta rp}\tilde{\mathcal{D}}_{\Delta rp}} \|DQD^{-1}\|_{\mathcal{K}_2\to\mathcal{K}_2}$; *thus Corollary G.5 specializes to Proposition 6.8.*

Bibliography

[1] M. Araki and Y. Ito, "Frequency-response of sampled-data systems I: open-loop considerations," *Proc. IFAC World Congress*, 1993.

[2] M. Araki, and Y. Ito, "Frequency-response of sampled-data systems II: closed-loop considerations," *Proc. IFAC World Congress*, 1993.

[3] B. Aupetit, *A Primer on Spectral Theory*, Springer-Verlag, New York, 1991.

[4] G.J. Balas, J.C. Doyle, K. Glover, A.K. Packard, and R.S. Smith, "The μ Analysis and Synthesis Toolbox," MathWorks and MUSYN, 1991.

[5] B. Bamieh, M.A. Dahleh, and J.B. Pearson, "Minimization of the $\mathcal{L}_\infty$-induced-norm for Sampled-data Systems," *IEEE Transactions on Automatic Control*, vol. 38, 717–732, 1993.

[6] B. Bamieh, and J.B. Pearson, "The $\mathcal{H}_2$ Problem for Sampled-data Systems," *Systems and Control Letters*, vol. 19, 1–12, 1992.

[7] B. Bamieh, and J.B. Pearson, "A General Framework for Linear Periodic Systems with Application to $\mathcal{H}_\infty$ Sampled-data Control," *IEEE Trans. Auto. Control*, vol. 37, 418–435, 1992.

[8] B. Bamieh, J.B. Pearson, B.A. Francis, and A. Tannenbaum, "A Lifting Technique for Linear Periodic Systems with Applications to Sampled-data Control," *Systems and Control Letters*, vol. 8, 1991.

[9] T. Basar, "Optimum $\mathcal{H}_\infty$ Designs Under Sampled State Measurements," *Systems and Control Letters*, vol. 16, 399–409, 1991.

[10] H. Bercovici, C. Foias, and A. Tannenbaum, "Structured Interpolation Theory," *Operator Theory Advances and Applications,* vol. 47, 195-220, 1990.

[11] B. Bollobas, *Linear Analysis*, Cambridge University Press, Cambridge, 1990.

[12] S. Boyd, L. Elgaoui, and V. Balakrishnan, *Linear matrix inequalities in systems and control theory*. SIAM, Philidelphia, 1994.

[13] S. Boyd, and C. Barratt, *Linear Control Design: Limits of Performance*, Prentice-Hall, 1991.

[14] S. Boyd, and Q. Yang, "Structured and Simultaneous Lyapunov Functions for System Stability Problems,"*Int. Journal of Control,* vol. 49, 2215–2240, 1989.

[15] T. Chen, and L. Qui, "$\mathcal{H}_\infty$ Design of General Multirate Sampled-data Control-Systems," *Automatica*, vol. 30, 1139–1152, 1994.

[16] T. Chen, "Control of Sampled-data Systems", Ph.D. Thesis, Electrical Engineering, University of Toronto, 1991.

[17] T. Chen, and B.A. Francis, "Stability of Sampled-data Feedback systems," *IEEE Trans. Auto. Control*, vol. 36, 50-58, 1991.

[18] T. Chen, and B.A. Francis, "Linear Time-varying $\mathcal{H}_2$-optimal control of sampled-data systems," *Automatica*, vol. 27, 963–974, 1991.

[19] J. Diestel, and J.J. Uhl, *Vector Measures*, Mathematical Surveys, no. 15, Providence, RI, AMS, 1977.

[20] J.C. Doyle, "Analysis of Feedback Systems with Structured Uncertainties," *IEE Proceedings*, vol. 129, 242–250, 1982.

[21] J.C. Doyle, "Structured Uncertainty in Control System Design," *Proc. CDC*, 1985.

[22] G.E. Dullerud, and K. Glover, "Analysis of Structured LTI Uncertainty in Sampled-data Systems," *Automatica*, vol. 31, 99–113, 1995.

[23] G.E. Dullerud, and K. Glover, "Robust Stabilization of Sampled-data Systems to Structured LTI Perturbations," *IEEE Trans. Auto. Control*, vol. 38, 1497–1508, 1993.

[24] G.E. Dullerud, and B.A. Francis, "$\mathcal{L}_1$ Design and Analysis in Sampled-data Systems," *IEEE Trans. Auto. Control*, vol. 37, 436–446, 1992.

[25] G.E. Dullerud, and K. Glover, "Necessary and Sufficient Conditions for Robust Stability of SISO Sampled-Data Systems to LTI Perturbations," *Proc. ACC*, 1992.

[26] L. Elsner, "On the Variation of the Spectra of Matrices," *Linear Algebra and its Applications*, 47, 127–138, 1982.

[27] C. Foias, and A.E. Frazho, *The Commutant Lifting Approach to Interpolation Problems*, Birkhäuser, Basel, 1990.

[28] C. Foias, A. Tannenbaum, and G. Zames, "On the $\mathcal{H}_\infty$-optimal sensitivity problem for systems with delays," *SIAM J. Control and Optimization*, vol. 25, 686–706, 1987.

[29] B.A. Francis, and T.T. Georgiou, "Stability theory for linear time-invariant plants with periodic digital controllers," *IEEE Trans. Auto. Control*, vol. 33, 820-832, 1988.

[30] B.A. Francis, *A Course in $\mathcal{H}_\infty$ Control Theory*, Springer-Verlag, New York, 1987.

[31] P.R. Halmos, *A Hilbert Space Problem Book*, Springer-Verlag, New York, 1982.

[32] S. Hara, and P.T. Kabamba, "Worst case analysis and design of sampled-data control systems," *Proc. CDC*, 1990.

[33] S. Hara, M. Nakajima, and P.T. Kabamba, "Robust Stabilization in Digital Control Systems," *Proc. MTNS*, 1991.

[34] E. Hille, *Functional Analysis and Semi-groups*, AMS, New York, 1948.

[35] K. Hoffman, *Banach Spaces of Analytic Functions,* Englewood-Cliffs, New Jersey, 1962.

[36] R.E. Kalman, and J.E. Bertram, "A Unified Approach to the Theory of Sampling," J. of Franklin Institute, vol. 267, no. 5, 1957.

[37] M. Khammash, "Necessary and sufficient conditions for the robustness of time-varying systems with applications to sampled-data systems," *IEEE Trans. Auto. Control*, vol. 38, 49–57, 1993.

[38] S. Lall, " Robust Control Synthesis in the Time Domain," Ph.D. Thesis, Engineering Department, Cambridge University, 1995.

[39] G.M.H. Leung, T.P. Perry, and B.A. Francis, "Performance Analysis of sampled-data control systems," *Automatica,* 699–704, 1991.

[40] D.G. Luenberger, *Linear and Nonlinear Programming*, Addision-Wesley, Reading, 1984.

[41] A. Megretski, "Neccessary and Sufficient Conditions of Stability: A Multiloop Generalization of the Circle Criterion," *IEEE Trans. Auto. Control*, vol. 38, 753-756, 1993.

[42] A. Megretski, and S. Treil, "Power Distribution Inequalities in Optimization and Robustness of Uncertain Systems," *Journal of Mathematical Systems, Estimation, and Control*, vol. 3, 301–319, 1993.

[43] R.A. Meyer, and C.S. Burrus, "A Unified Analysis of Multirate and Periodically Time-Varying Filters," *IEEE Trans. on Cir. and Sys.*, vol. CAS-22, 162–168, 1975.

[44] Y.E. Nesterov and A. Nemirovsky, *Interior Point Polynomial Methods in Convex Programming,* SIAM, Philidelphia, 1994.

[45] Y. Oishi, *A Study on Robust Control of Sampled-data Systems*, Master of Engineering Thesis, University of Tokyo, 1993.

[46] E.E. Osborne, "On Preconditioning of Matrices," *J. Assoc. Comp. Mach.*, vol. 7, 338–345, 1960.

[47] A. Packard, "What's new with μ: Structured Uncertainty in Multivariable Control," Ph.D. Thesis, Mechanical Engineering, University of California, Berkeley, 1988.

[48] F. Paganini, and J.C. Doyle, "Analysis of Implicitly Defined Systems," *Proc. CDC*, 1994.

[49] A. Packard, and J.C. Doyle, "The Complex Structured Singular Value," *Automatica*, Vol. 29, No. 1, 71–109, 1993.

[50] K. Poolla, and A. Tikku, "Robust Performance Against Slowly-Varying Structured Perturbations," *CDC Proceedings*, 1993.

[51] L. Qui, and T. Chen, "$\mathcal{H}_2$ Optimal Design of Multirate Sampled-data Systems," vol. 39, 2506–2511, 1994.

[52] S. Rangan, and K. Poolla, "Time-domain Sampled-data Model Validation", *Proc. ACC*, 1995.

[53] R.M. Redheffer, "On a certain linear fractional transformation,' J. Math. and Physics, vol. 39, 269–286, 1960.

[54] R.T. Rockafellar, *Convex Analysis*, Princeton University Press, Princeton, New Jersey, 1970.

[55] W. Rudin, "Boundary Values of Continuous Analytic Functions", *AMS Proceedings*, vol. 7, 809-811, 1956.

[56] M.G. Safonov, "Tight Bounds on the response of multivariable systems with component uncertainty," *Proc. Allerton*, 451–460, 1978.

[57] I.W. Sandberg, "An Observation Concerning the Application of the Contraction Mapping Fixed-point Theorem and a Result Concerning the Norm-boundedness of Solutions of Nonlinear Functional Equations," *Bell Systems Tech. J.*, vol. 44, 1809–1812, 1965.

[58] J. Shamma, "Robust Stability with Time-Varying Structured Uncertainty," *IEEE Trans. Auto. Control*, vol. 39, 714–724, 1994.

[59] N. Sivashankar and P.P. Khargonekar, "Robust Stability and Performance Analysis of Sampled-Data Systems," *IEEE Trans. Auto. Control*, vol. 38, 58–69, 1993.

[60] N. Sivashankar and P.P. Khargonekar, "$\mathcal{H}_2$ Optimal Control for Sampled-data Systems," *Systems and Control Letters*, vol. 17, 425–436, 1991.

[61] R. Smith, and G.E. Dullerud, "Validation of Continuous-time Control Models by Finite Experimental Data", *Proc. ACC*, 1995.

[62] W. Sun, K.M. Nagpal, and P.P. Khargonekar, "H_∞ Control and Filtering for Sampled-data Systems," *IEEE Trans. Auto. Control*, vol. 38, 1162–1175, 1993.

[63] B. Sz.-Nagy, and C. Foias, *Harmonic Analysis of Operators on Hilbert Space*, North-Holland, Amsterdam, 1970.

[64] P.M. Thompson, R.L. Daily and J.C. Doyle, "New conic sectors for sampled-data system feedback systems," *Systems and Control Letters*, vol. 7, 395-404, 1986.

[65] P.M. Thompson, G. Stein and M. Athans, " Conic Sectors for Sampled-Data Feedback Systems", *Systems and Control Letters*, vol. 3, 77-82, 1983.

[66] H.T. Toivonen, "Sampled-data control of continuous-time systems with an H_∞ optimality criterion," Rep. 90-1, Dept. Chemical Eng., Abo Akademi, Finland, Jan. 1990.

[67] C. Van Loan, "Computing integrals involving the matrix exponential," *IEEE Trans. Auto. Control*, 23, 395–404, 1978.

[68] J. Weidmann, *Linear Operators on Hilbert Spaces*, Springer-Verlag, New York, 1980.

[69] Y. Yamamoto and P.P. Khargonekar, "Frequency Response of Sampled-data Systems," *Technical Report* Dept. of Physics and Applied Mathematics, Kyoto Univ., 1993, *submitted to IEEE TAC.*

[70] Y. Yamamoto, "A new approach to sampled-data control systems — A function space viewpoint with applications to tracking problems," *Proc. CDC*, 1882–1887, 1990.

[71] N. Young, *An Introduction to Hilbert Space*, Cambridge University Press, Cambridge, 1988.

[72] P.M. Young, "Robustness with Parametric and Dynamic Uncertainty," Ph.D. Thesis, Electrical Engineering, California Institute of Technology, 1993.

[73] G. Zames, "Feedback and Optimal Sensitivity: Model Reference Transformations, Multiplicative Seminorms and Approximative Inverses", *IEEE Trans. Auto. Control* , vol. 26, 301-320, 1981.

[74] G. Zames, "On the Input-Output Stability of Time-varying Nonlinear Feedback Systems. Part I: Conditions using Concepts of Loop Gain, Conicity, and Positivity," *IEEE Trans. Auto. Control* , vol. 11, 228-238, 1966.

[75] K. Zhou and P.P. Khargonekar, "On the weighted sensitivity minimization problem for delay systems," *Systems and Control Letters*, vol. 8, 307–312, 1987.

[76] A. Zygmund, *Trigonometric Series*, Cambridge University Press, Cambridge, 1968.